VG
(2014)

D1431034

Optimal Supply Chain Management in Oil, Gas, and Power Generation

OPTIMAL
SUPPLY CHAIN
MANAGEMENT
in Oil, Gas, and Power Generation

DAVID JACOBY

Copyright© 2012 by
PennWell Corporation
1421 South Sheridan Road
Tulsa, Oklahoma 74112-6600 USA

800.752.9764
+1.918.831.9421
sales@pennwell.com
www.pennwellbooks.com
www.pennwell.com

Marketing Manager: Amanda Alvarez
National Account Executive: Barbara McGee

Director: Mary McGee
Managing Editor: Stephen Hill
Production Manager: Sheila Brock
Production Editor: Tony Quinn
Book and Cover Designer: Susan E. Ormston

Library of Congress Cataloging-in-Publication Data

Jacoby, David.
  Optimal supply chain management in oil, gas, and power generation / David Jacoby.
      p. cm.
  Includes bibliographical references and index.
  ISBN 978-1-59370-292-2
  1. Petroleum industry and trade--Management. 2. Energy industries--Management. 3. Business logistics. I. Title.
  HD9560.5.J258 2012
  622'.3380687--dc23
                                                    2012024727

CONTENTS

Part 1

Part 2

Illustrations

Figures

Tables

Foreword

by Dr. Hamza Abada
Business Support Advisor, RasGas
Doha, Qatar

Supply chain management tools and techniques have been adopted and applied in many industries and businesses to varying levels. In some industries, the full supply chain process is followed, and the supply chain manager is one of the main decision-makers in the organization; in others, the supply chain is only a procurement function, where a number of buyers are ensuring that goods and services are ordered and delivered. Regardless, the core role of supply chain management is to align procurement and supply with the strategic goals of the organization and to ensure that these goals are achieved in the most effective and efficient way.

The challenges in the oil and gas industry involve a continuous search for new ways to reduce time to market, streamline processes, make collaboration easier, increase revenues, and cut costs—all while delivering projects and solutions in time. However, there may be no other industry today that demands a more diverse set of human, political, mechanical, and technological capabilities than oil and gas exploration and production. The technical complexities, the high risks and the impact of these risks, the size of investments, and the continuous rising demand for energy make supply chain management in the oil and gas industry a critical activity for the success of any project in this field.

The breadth of the oil, gas, and power industries requires supply chain practitioners with diverse knowledge—not only on the business side but also on the technical side—to enable them to procure the right tools and services at the most competitive price and quality with minimum risk exposure. Technical managers and engineers require knowledge of the available supply strategies to make sound decisions regarding the technology they should use and the equipment they should request.

This book provides well-defined tools for managing today's risk and cost pressures in oil, gas, and power supply chains. These industries are growing so rapidly that conventional methods have quickly become obsolete, and projects are of such a large scale that risk sharing between owners, operators, contractors, and suppliers needs to be addressed consistently and safely. Large economic decisions have historically been made by gut feeling, but as risks and opportunities increase, our intuition needs to be supplemented by rigorous logic, reliable data, and robust analysis.

The book differs from other supply chain books in two important ways. First, it focuses on unique characteristics and requirements of the oil, gas, and power industries, offering hundreds of examples and case studies to ensure applicability and relevance to various operations. Second, it provides methodologies for making hard business decisions that involve risk, for which there have been missing or inadequate frameworks for measuring and managing that risk (hence, the reliance on instinct and intuition in the past).

The perspective and needs of producers (owners and/or operators) is emphasized. The coverage of shipping operations (in particular, LNG, with less attention to crude) is limited to a brief overview of transport management, since these are specialized areas that may not be relevant to a broad audience of readers.

Traditional logistics functions such as procurement, logistics, and materials management are discussed in detail. In addition, health, safety, environmental, and other regulatory areas are covered that have substantial interfaces with supply chain management decisions. Information technology systems (e.g., warehouse management systems, transportation management systems, and e-procurement) are discussed only in brief, as means to an end, not supply chain management ends in themselves.

The discussion of key processes covers capital expenditure planning, operating cost reduction, asset management, and risk management. Transactional issues, such as electronic data interchange and paperless work flow, are excluded to focus instead on critical supply chain projects involving capital procurement and large risk and the supply chains that support those projects.

This book will serve technical managers and engineers as a reference and guide to various supply strategies, while supply chain professionals will find it to be a comprehensive guide to most oil and gas activities. The real-life case examples provided in this book are refreshing and should provide a welcome resource for readers interested in supply in the oil and gas industry.

Preface

The Growing Importance of Supply Chain Management in Oil, Gas, and Power

In the mid-1800s in the United States, drilling and production was concentrated over only a few geographical areas, and refining was done at small-scale refineries. In the 1930s in Saudi Arabia, workers crushed minerals by hand to create drilling mud. The California-Arabian Standard Oil Company initially employed Bedouin tribes to guard their fields and supply lines, transitioning as its operations expanded to government police and eventually to private security. In 1938, before the advent of domestic refining capacity and regional pipelines, Saudi Arabia exported crude by barge to Bahrain. By the 1940s, tanker trucks were transporting oil from the production site to refineries. Each held 40–50 barrels, compared to 120–215 barrels for tanker trucks today. Pipelines were also used, but they were not the main mode of transportation. From the refinery, finished products were mostly shipped by rail.

Higher well production and improved technology eventually made pipelines the most efficient means of long-distance oil transport. Developers pioneered stronger, more durable materials and more reliable construction methods. Operators also began using electronic flow measurement, developed in-line inspection (smart pigging) and satellite corrosion monitoring, and implemented supervisory control and data acquisition systems to manage the pipelines.

The emergence of megaprojects

The current boom in investment and production—both in oil and gas and in power—has transformed supply chain management into a different game, with high stakes. Capital expenditures within the oil, gas, and power generation industries have grown at more than 15% per year since 2004, with the exception of 2009 and 2010, the years following the financial crisis and of the Macondo disaster in the Gulf of Mexico that caused a temporary shutdown of deepwater drilling.[1] Current investment in infrastructure development is impressive by historical standards. Saudi Arabia's power projects alone represent about $130 billion in total investment between 2010 and 2015.

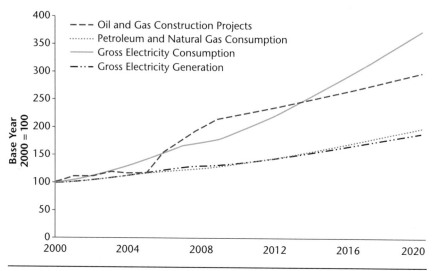

Fig. P–1. Evolution of growth in oil, gas, and electricity markets[2]

Megaprojects are underway in various countries including the United Arab Emirates, Kuwait, Qatar, and Egypt. In the Emirates, "Abu Dhabi is planning three large-scale power projects, and the government awarded a $20.4 billion contract for four nuclear reactors to a consortium led by Korea Electric Power (KEPCO) in 2009. The first reactor is expected to come onstream in 2017, and all four will be producing electricity by 2020. Kuwait is conducting the feasibility study for a $20 billion nuclear power project with France. Qatar has been busy building more conventional power infrastructure, with a $5 billion upgrade to its power transmission network including 54 new substations, a $5 billion coal-fired power plant in Raysut, and a $5 billion solar power capability. Qatar Petroleum's Ras Laffan Independent Power Project (IPP) is budgeted at about $4 billion (not counting Ras Laffan C), and Kahramaa's West Coast Independent Water and Power Plant (IWPP) will have cost about $3 billion."[3] In addition to projects in specific countries, the Gulf Cooperation Council (GCC) power grid project will link the GCC's grid, creating the opportunity for a massive smart grid, and Saudi Arabia and Egypt are working on an $8 billion, 1,370-kilometer interconnection project using overhead transmission lines.

As these and other megaprojects move through their planning, construction, and operating phases, the number of actors required to realize them has transformed supply chain management into a critical tool. All parties have to manage cost and risk up and down the value chain to meet schedule and budget projections. In the financing and planning stages, the demand or revenue forecast could be wrong.[4] At this stage, owners and financiers may also make erroneous assumptions about the cost of disposal of used fuel and radioactive waste. In the construction stage, estimates of materials and services costs may be off, and so may assumptions about exchange rates, bottlenecks at engineering, procurement, and construction contractors and subcontractors and about the extent and cost of design changes. In the operation phase, volume, seasonality, and growth curves may not be on target. Volatile and escalating electricity tariffs and fuel costs are hard to forecast. The rate of return on invested working capital and earmarked funds may be more or less than anticipated. Government regulations, such as the laws governing investment and repatriation of profits, and institutional frameworks, such as the mechanics and limits for trading electricity on the open market, may change.

New global challenges and their supply chain implications

Supply chain risks are further exacerbated by nontraditional challenges: raw material depletion, technological complexity, resource nationalism, environmental pressures, and globalization are making coordination up and down the supply chain more critical than ever.

Resource depletion has raised the cost and created shortages of critical raw materials, requiring a search for new technologies and substitute products. According to the global management consulting firm McKinsey, the 21st century needs a resource revolution akin to the labor revolution of the 20th century—during which labor productivity soared owing to information technology—to fulfill the rapidly growing needs of our global population.[5]

Water resource depletion is affecting the market for steel, which is becoming ever more critical in the oil and gas sector because of its use in huge offshore platforms and associated pipelines. The supply of iron ore is crucial to steel production and in turn relies heavily on water to reduce the ore. McKinsey notes that almost 40% of iron ore mines are in areas with moderate to high water scarcity, and a large amount of steel is produced in places where water is relatively scarce.[6]

Scarce mineral resources are driving up costs for drilling fluids and bits. The gradual depletion of barite, which is used as a weighting agent, is driving costs up dramatically. The world's identified barite reserves will be depleted in little more than a decade.[7] In addition, China, the largest barite exporting country, vowed in its latest five-year plan to restrict the environmental impact of mining for minerals such as barite and to control their export more tightly, which could drive barite prices up. Major suppliers are scouring the globe to develop alternative barite sources. Tungsten, a material used to harden drill bits and increase their durability, was in tight supply in 2008 when rig counts rose rapidly, creating both price inflation and bottlenecks. Similar to barite, the mineral is mostly mined in the interior of China and is vulnerable to supply disruptions from the country's ongoing mining-sector reforms.

Increasing technological complexity related to depletion of petroleum reserves requires fewer and larger suppliers that have the capacity, the financial resources, and the broad technical skills needed to develop the extraordinary new technology at the edge of science and to support megaprojects. Offshore and deepwater projects are unlocking huge reserves but require capital investment many times that of traditional, onshore developments. Unconventional gas projects across the United States, oil sands and heavy oil projects in Canada, drilling in the Arctic, and ultra-deepwater discoveries off the coasts of Brazil and Angola are increasing net available energy resources, but they require unprecedented capital and technology. Major research and development efforts are underway to produce heavy oil economically and to treat oil with high dissolved hydrogen sulfide content that will allow production to continue in fields across Asia and the Middle East that would have otherwise been plugged. Finally, Australian offshore natural gas deposits, such as the Icthys gas field, are extending the limits of science on a daily basis, but the conditions are so harsh that only the majors need apply.

Enhanced oil recovery is being combined with carbon capture and sequestration (CCS), which requires ultralarge compressors that only a handful of suppliers have the technology and scale to produce. Enhanced oil recovery, the method of increasing oil yield from older oil wells by injecting gases or chemicals to increase the pressure and output, has been around for years (e.g., Norway began injecting carbon dioxide into the North Sea in 1996), but the search for carbon dioxide and the injection thereof has been done on a well-by-well basis. Connection of existing sources of carbon dioxide to oil fields benefits both power generators and oil drillers: power generators get lower emissions, and oil drillers can increase recovery rates by up to 75% in some cases.[8] However, the economics of CCS are far from proven, which leaves large companies to manage the initial projects.

The application of information technology to obtain real-time intelligence about downhole conditions is helping operators to minimize fluid loss, manage flows within the reservoir, and reduce the need for interventions, thereby maximizing production. However, given the opportunity cost of well production, operators trust only suppliers with proven solutions.

Resource nationalism is propelling a trend toward more local content, which requires developing suppliers that otherwise may not have been prequalified. According to one source, since 1970 national oil companies (NOCs) have increased their control over the world's hydrocarbon reserves from 15% to 85%,[9] while international oil companies (IOCs) have shrunk from 85% to 15%. This transition has occurred through the changing nature of production sharing agreements, which are increasingly oriented to provide NOCs with title to hydrocarbons, control over operations, upside risk on oil price and volume movements, and a greater share of the profit upside. As NOCs have taken more control of their countries' resources, they have increasingly required foreign contractors to use a greater share of domestically produced goods and services. Countries actively setting local content levels include Kazakhstan, China, Brazil, and Azerbaijan.

Growing public environmental conscience is making tight process integration with suppliers a requirement for survival. Two high-profile incidents in 2010—the Macondo disaster and a public outcry over possible health risks posed by the fracturing fluids used heavily in shale plays—intensified scrutiny of safety and security in the oil and gas supply chain. The first major deadline for registration of chemical substances affected by the European Union's regulation on the Registration, Evaluation, and Authorisation of Chemicals (REACH) kept the public eye focused on environmental impact, and in the United States an Environmental Protection Agency subpoena invoked the Toxic Substances Control Act of 1976, the Clean Water Act of 1972, and the Resource Conservation and Recovery Act of 1976 to force public disclosure of certain formulations that would have previously been considered trade secrets or proprietary ingredients. As a result, new competitors have been able to enter the market, but in the longer term, innovation may be stifled; the ultimate impact on industry structure remains to be seen.

The increasingly global nature of oil and gas operations has driven suppliers to expand their manufacturing and sales networks worldwide. Most floating production, storage, and off-loading (FPSO) projects are in emerging countries, which are forecast to grow at two to three times the rate of the developed countries.[10] Of 194 global FPSO projects, 115 are in Brazil, Africa, or Southeast Asia. Of these, 51% are in deepwater (1,000–1,500 meters) or ultra-deepwater (>1,500 meters).[11] Furthermore, India is emerging as a substantial oil and gas player because of large-scale drilling projects at state-run oil producers Oil and Natural Gas Company (ONGC), Indian Oil Company (IOC), and Gas Authority of India (GAIL). Suppliers have been aggressively migrating to the growth areas, notably China, India, and Malaysia. Moreover, the regional basis of oil demand has shifted in recent years (2008–2012)—with patterns of trade and transportation moving from countries in the Organisation for Economic Co-operation and Development (OECD) to non-OECD countries and from West to East. According to 2012 International Energy Agency projections, demand for crude oil has decreased 2.2% annually in Europe since 2008 and 0.9% in North America. Conversely, demand for crude oil from non-OECD countries in Asia has grown 4.7% annually over the same period. As a result, Asia now makes up 32% of global crude oil demand, as compared to 29% in 2008. Meanwhile, the share of global demand has dropped from 18% to 16% in OECD Europe and from 28% to 26% in North America.[12]

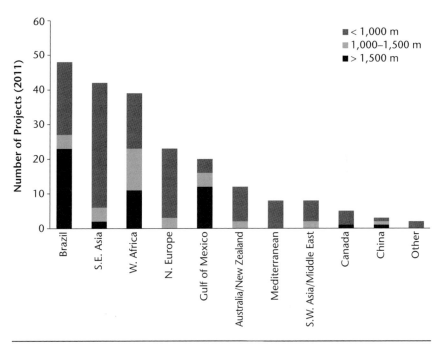

Fig. P–2. Number of FPSO projects by country or region (*Source:* Floating Production Systems: Assessment of the Outlook for FPSOs, Semis, TLPs, Spars and FSOs. International Maritime Associates, Inc., 2012.)

The globalization of the industry is forcing suppliers to respond with global service and more robust international logistics capabilities. Schlumberger now has 23 service centers around the world, many with field staff providing installation, troubleshooting, preventive maintenance, and repair. Baker Hughes has opened a service and operations base in Saudi Arabia.[13] Dresser-Rand plans to open 20 or more service and support centers to accelerate response to international opportunities.[14] Pump manufacturer Ruhrpumpen established service centers in Mexico, the United States, Germany, Egypt, Canada, and Argentina. As suppliers globalize, the cost of shipping internationally is forcing them to reevaluate their supply chains, sometimes replacing their raw material sources and reconfiguring their intermediate processing activities and locations. While this presents opportunities for local suppliers, it can also threaten them. In Saudi Arabia, local manufacturers like Saudi Cement Company and Saudi Yamama Cement have traditionally dominated the cement supply market. However, global leaders such as Lafarge and Cemex have set up plants to capture a share of the growing regional construction market in response to the Saudi Arabian government's $400 billion infrastructure plan.

Purpose and Organization of This Book

The book provides a toolbox for large-scale capital expenditure decision making and for transforming capital and operating expenditure to exert a visible financial impact at the enterprise level in oil, gas, and power companies. By using the supply chain risk management decision analysis tools in this book, operators can increase economic value added (EVA) by 3.8% (higher on greenfield capital projects, lower on existing infrastructures) while enhancing stewardship to safety and the environment. This is based on a 13% reduction in capital costs, a 9.8% increase in total annual sales revenue as a result of debottlenecking and increased throughput capacity, and an average 1.0% reduction in total operating costs.

The book is designed to be read in two passes. First, read the introduction and all of Part 1. This covers the overall concepts and principles that are relevant to the oil, gas, and power industries. Then, read the chapter in Part 2 that applies to your business—upstream, midstream, or downstream oil and gas or power generation.

Part 1

The initial chapter on capital expenditure and supply chain planning lays out concepts and principles for designing supply chains, architecting supplier relationships, managing contract risk, and engineering and constructing projects. It provides tools for managing the challenge of making commitments prior to final investment decision and describes the theory and application of options as they relate to project size, project life, and technology/ product mix. This chapter also elaborates on multiple methods for hedging the risk of material and service unavailability and price volatility. Finally, it addresses the complex issue of how to define the buy when structuring project work—that is, whether to buy solutions or independent products and services, how to do a proper analysis of total cost of ownership, which project governance mode to select, which activities to insource and which to outsource, and how to determine the optimal contract term.

Chapter 2 includes a section on developing bid slates and explains the concept of category management. Methods are illustrated to determine the optimal number of suppliers and qualify suppliers once they are identified. A section on structuring partner relationships explains how to structure alliances, including whether to establish a joint venture and whether to take equity in a partner. Next, a section on build-own-operate choices defines a mutually exclusive and collectively exhaustive set of alternatives for how much to risk to bear and how much to shift to other parties. Finally, a section on managing tendering procedures provides guidance and tools for tendering products under development but not yet commercialized, tendering combined purchase and operating/maintenance agreements, tendering to local sources where local content regulations apply, and using auctions.

Chapter 3 is divided into three sections. The first section, on maintaining complex systems, provides frameworks and proven processes and measurements for managing asset productivity, total cost of ownership/life-cycle cost, the cost of quality, throughput and debottlenecking, preventive and predictive maintenance, and standardization of equipment, services, and processes. The second section, on achieving continuous cost reduction, explains the concepts of lean as they relate to inventory management, transportation management, outsourcing logistics, and total supply chain activities (third- and fourth-party logistics service providers), and how to engage suppliers in performance improvement initiatives. The third section, on logistics, inventory, and materials management, explains how to manage capital spare parts differently from maintenance, repair, and operating supplies and how to establish consignment, vendor-managed inventory, and other inventory programs.

Chapter 4 lays out risk management standards, enterprise risk management frameworks and methods, and operational risk-mitigation processes and methods. It also clarifies supply chain's role in reducing environmental footprint, and it includes an explanation of the key elements of the Carbon Disclosure Project.

Part 2

Part 2 is divided into four chapters providing examples relating to a particular industry segment: upstream oil and gas; midstream oil and gas; downstream oil and gas; and power generation. Each chapter follows a common structure but treats a variety of different subtopics to the extent that they are important in each segment. Each chapter consists of at least four common topics:

- Introduction, describing unique supply chain characteristics
- Project risk mitigation, including best practices and contingency planning
- Engineering and procurement of equipment and services at minimum cost and risk
- Operations and maintenance cost reduction

A conclusion lays out key success factors for achieving the benefits articulated in the book, and suggests areas for future research.

The book is designed as a reference resource, so abundant endnotes allow for follow-up research or investigation into specific topics. Many of the referenced documents are previous publications of mine. They are all available at www.bostonstrategies.com. Furthermore, the book has four appendixes:

- Appendix A is a study showing how the lack of effective supply chain coordination results in higher costs for upstream and downstream producers and their suppliers (i.e., bullwhip).
- Appendix B gives a useful reference list for categorizing purchased equipment and services.
- Appendix C is a glossary defining acronyms and abbreviations and explaining frequently used terms as they

apply to oil, gas, and power generation companies. The definitions are intentionally focused on the applications described in this book, so they are specific to the context of supply chain management; thus, they may differ from the standard definitions in more generic sources. They are also intended to be short, to the point, and practical, rather than comprehensive and academic.

• Appendix D provides a list of additional resources, for readers who want to dive further into specific topics.

Finally, readers are invited to send any suggestions for improvements to future editions to me at djacoby@bostonstrategies.com.

Notes

1 Boston Strategies International analysis based on HPI Market Data 2010, HPI Construction Boxscore, and Economist Intelligence Unit Data Tool.

2 HPI Market Data Book 2010. Hydrocarbon Processing. Houston: Gulf Publishing Company, p. 6. Data from Economist Intelligence Unit, London, www.eiu.com.

3 Jacoby, David. 2010. Balancing economic risks: tips for a well-structured deal. *Middle East Energy* (September): 5.

4 Jacoby, David. 2009. *Guide to Supply Chain Management: How Getting It Right Boosts Corporate Performance*. Economist Books. New York: Bloomberg.

5 Dobbs, Richard, Jeremy Oppenheim, and Fraser Thompson. 2012. Mobilizing for a resource revolution. McKinsey Global Institute, Sustainability and Resource Productivity Practice, in *McKinsey Quarterly—Energy, Resources, Material*, January, p. 1.

6 McKinsey. Dependencies and regulatory risks.

7 Based on an extrapolation of data from the U.S. Geological Survey, Mineral Commodity Summaries, January 2006.

8 Wahbi, Salah Hassan. 2010. Sudan's growing exploration and development. Paper presented at the National Oil Conference, London, June.

9 Burdis, Ian. 2010. Strategies and practices to realize the maximum potential of NOC's and IOC's. Paper presented to NOC Congress, June 23, p. 5.

10 Jacoby, David. 2011. Uncovering economic and supply chain success in the new emerging economies. Paper presented to APICS International Conference, Pittsburgh, October 24.

11 Floating Production Systems: Assessment of the Outlook for FPSOs, Semis, TLPs, Spars and FSOs. International Maritime Associates, Inc., 2012.

xxvi | Optimal Supply Chain Management in Oil, Gas, and Power Generation

12 International Energy Agency. 2012. *Oil Market Report*. Paris: International Energy Agency.

13 Baker Hughes. 2010. Baker Hughes opens new operations base in Saudi Arabia. Baker Hughes press release, April 5. http://www.bakerhughes. com/assets/media/pressreleases/4bbb9e8b1772316225000001/file/bhi_ news_2010_4_5_product_lines.pdf.pdf&fs=11243.

14 Robb, Drew. 2011. Compressor maintenance trends: Modularity, remote monitoring and outsourcing are key. *Turbomachinery International*. September/ October. http://www.turbomachinerymag.com/sub/2011/SeptOct2011- CoverStory.pdf.

Acknowledgments

This book is dedicated to the distinguished leaders who are boldly improving the efficiency and performance of our oil, gas, and power supply chains, especially Alfred Kruijer from Shell, Stephen Turnipseed from Chevron, Nayef Al-Hajri and Khaled Baradi from Qatar Fuel, Gerald Sheils and Phil Roberts from Saudi Aramco, Magnus Öhrman from Vattenfall, Suresh Nair from Bharat Petroleum, Gary Wawak from Motiva, Art Soucy from Baker Hughes, Dave Cox from GE Oil & Gas, Peter Krieger from Freudenberg Oil and Gas, Åsmund Mandal from FMC Kongsberg, Somchai Kooyai from PTT Group, Eric Boyles from Elliott Group, and James MacLean from Geoforce.

I would also like to acknowledge the American Petroleum Institute (API), the International Organization for Standardization (ISO), the International Maritime Organization, the Committee of Sponsoring Organizations of the Treadway Commission, and Det Norske Veritas for the leadership they have provided in risk management, operations excellence, and best practices. In addition, several professional associations have developed extensive and useful bodies of knowledge in supply chain and operations research, including (but not limited to) the Association for Operations Management, the Institute for Supply Management, and the Council of Supply Chain Management Professionals.

Special thanks go to the consultants and partners at Boston Strategies International who supported the effort to write this book despite their already full workload. Erik Halbert helped prepare the initial outline, contributed sections, and carefully reviewed drafts; Yingying Gu and Fonahanmioluwa Adesola Osunloye checked facts and attributions and logged permissions to reprint charts and tables; Alok Gupta, Aditya Sharma, Sidharth Sharma, Gautam Marwaha, and Samir Canning researched and organized examples and case studies; and Nicolai Jakobsen and others occasionally absorbed the workload of those who were drawn off client work to support this book project. Thanks are also due to Dennis O'Dea for his careful review of the draft manuscript.

Thank you also to PennWell, and to Stephen Hill in particular, for nurturing the development of the project from concept through to outline and full manuscript. Without their faith and experienced guidance, this book would not have come to fruition.

Finally, and although they would undoubtedly state for the record that they never agreed to this book project (my second), I thank my wife, Jessica, and my children, Weston, Brent, and Camille, for indulging my passion in various work projects including this book, which frequently involve travel, time away from family, or both.

David Jacoby
Boston, April 2012

INTRODUCTION

How Supply Chain Management in Oil, Gas, and Power is Different

Most people's understanding of supply chain management stems from the consumer products industry, in which thousands of stock-keeping units (SKUs) of fast-moving consumer goods flow through distribution centers and move from pallets onto retail shelves. While this characterizes *some* logistics flows that occur in oil, gas, and power, supply chain management in these industries is different in several important ways.

Supply chain management in oil, gas, and power more closely resembles supply chain management in the process industries (i.e., those with continuous production operations). Even so, it is much more complex than in low-value process industries such as paper and cement. While it more closely resembles supply chain management in high-value process industries such as petrochemicals and pharmaceuticals, it is different enough from those industries to have its own body of knowledge.

Technology, and often chemistry, affects every decision, from network design to procurement, installation, and logistics. In the offshore segment, logistics draw heavily on the experiences of the maritime industry, with purpose-built vessels to tap hydrocarbon reserves farther offshore and in deeper waters. This complexity manifests itself in most segments of the business—for example,

- In upstream oil and gas, replacing a worn part on a subsea wellhead is more complex than stocking an item in a bin. It involves a heavy dose of chemistry, physics, fluid mechanics, and possibly a remote-controlled underwater vehicle.

- In midstream oil and gas, shipping of liquefied natural gas (LNG) involves a deep knowledge of cryogenics. Natural gas flow and storage are affected by temperature and pressure changes (which depend on heat transfer, changes in viscosity and surface tension, and erosion and corrosion problems). Natural gas flow is ensured by minimizing corrosion, preventing hydrate formation conditions, predicting the effectiveness of inhibitors, preventing wax deposition that may impact pressure and flow, avoiding slugging (i.e., liquid buildup that prevents gas from flowing), adjusting the number and pressure of the producing wells, and closely monitoring shutdowns and restarts.[1]

- In downstream oil and gas, the choice of processing routes (inputs, intermediate outputs, inter-plant transport, and final product choices) requires multirefinery production optimization involving permutations of a large number of possible chemical grades at each echelon of the supply chain.

- In the power industry, advances in materials science are allowing the temperature inside gas and steam turbines to exceed 2,200°F, which is critical to delivering more power.

Higher profit margins, at least in the oil and gas industry, shift supply chain management priorities from inventory management, which is often the main focus for fast-moving goods, toward reliability, safety, asset management (risk, utilization, and productivity), and life-cycle cost. Highly engineered equipment operates in complex and customized systems, and the question of how much additional engineering and customization is worth is perennial. Technology and supplier selection is done on the basis of life-cycle cost, and in the case of new technological platforms, there is little or no performance history on which to make accurate calculations.

The prohibitive cost of downtime makes reliability and field responsiveness critical. Since many assets are remote and/or offshore, the question of how to accurately value the opportunity cost of lost production is a recurring one, as well as the issue of how to factor those costs into inventory parameters for items like capital spares.

Long investment cycles mean decision-making must account for uncertainty. Construction decisions are made years in advance, and actual construction projects usually last for years. Concessions are frequently awarded for decades, which increases the importance of risk management and analytical risk management tools.

Dangerous conditions—for example, subsea and deepwater offshore—present large potential safety hazards that result in extraordinary price premia for safety and process reliability. This highlights another trade-off that is less common in other industries: How much is a "safe" item worth above and beyond the cheap version that has a chance of failure? (E.g., how much more should an explosion-proof light fixture cost than a regular one?)

Finally, the high public visibility of accidents and environmental problems elevates the importance of risk management, including supply chain risk. Even if the risk of serious problems is low, the possibility of having to face public scrutiny over a decision to save a small amount of time or money often makes managers wary of aggressive cost-cutting projects.

The bottom line is that the oil, gas, and power industries depend more on asset management and reliability than they do on velocity, customization, flexibility, and many of the supply chain principles and disciplines that have become buzzwords in the broader supply chain community. This means that supply chain practitioners will tend to work more on the "left side" of the supply chain matrix (see fig. I–1)—representing the methods designed to contain capex and operating cost—than on the right, which is principally concerned with using yield pricing strategies and high-touch customer interactions to improve profitability.

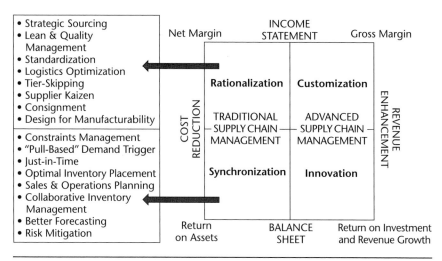

Fig. I–1. Relevant subset of supply chain management strategies for oil, gas, and power (*Source:* after Jacoby, 2009, p. 55.)

Overall Objectives, Key Business Processes, and Targets

The set of practices described in this book can add 3.4% to *economic value added* (EVA) and reduce downside operational risk by conforming to industry-accepted best practices. However, they need to be actively adapted, implemented, and governed.

Supply chain objectives

A well-articulated supply chain management mission that aligns with corporate goals will preempt many questions and debates. Oil, gas, and power operators fall into Types I and II supply chains (see fig. I–2). Upstream activities fall into Type I while downstream activities often fall into Type II.

There are four generic supply chain strategies, and the most appropriate one depends on the industry and line of business. The part of the supply chain strategy that is "given" is dictated by the business's place in the value chain (table I–1).

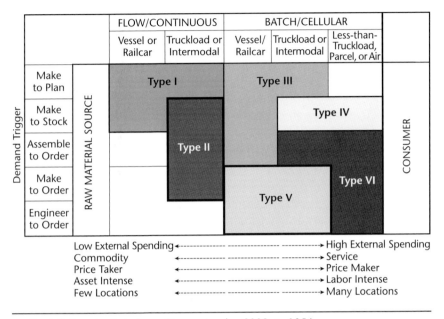

Fig. I–2. Supply chain types (*Source:* Jacoby, 2009, p. 105.)

Table I–1. Correlation between value chain role and supply chain strategy (*Source:* Jacoby, 2009, p. 53.)

Type of Business	Rationalization	Synchronization	Customization	Innovation
Extraction	Positive	Negative	Negative	Negative
Process (continuous) manufacturing	Positive	Negative	Negative	Negative
Batch manufacturing	Negative	Positive	Negative	Negative
Make-to-order manufacturing	Negative	Positive	Positive	Positive
Distribution	Negative	Negative	Positive	Positive
Reselling	Negative	Negative	Negative	Positive

Two generic supply chain strategies are most applicable to oil, gas, and power companies: *rationalization* and *synchronization*.

- Rationalization is excellence in managing operating costs through supply chain management to achieve cost leadership and greater profitability than competitors. Rationalization focuses on operating expense management rather than asset management, in particular for companies that are driven by quarterly earnings. Rationalization includes supply chain processes such as SKU rationalization (this refers to the reduction in the number of SKUs rather than a reduction of inventory, although the former can cause the latter), Kaizen, and value engineering and procurement activities such as sourcing (performance management, supplier selection, etc.), production, and facilities management.

- Synchronization is excellence in achieving reliable and flawless supply chain execution (i.e., right product at the right place at the right time) so as to be able to produce the same volume of output with less fixed assets (production capacity) and working capital (inventory) than competitors, more reliably, and with less variability and hence reduced risk. Synchronization includes processes such as inventory management, maintenance, routing and scheduling, demand planning, and risk management, as well as improvement programs such as product life-cycle management, Six Sigma, design for assembly, standardization, and collaborative forecasting, planning, and replenishment.

These two dominant strategies leverage four common cost drivers:

- *Standardization.* This minimizes the number of items, parts, components, and technological platforms in service, to maximize repetition within manufacturing and aftermarket service and to pool inventory, including spare parts.

- *Economies of scale.* These reduce cost owing to volume purchase and amortized fixed costs across greater output (including long-term contracts).

- *Competitive negotiation.* Bidding usually reduces prices as suppliers reduce their margin for an equivalent deliverable to win the order.

- *Learning-curve effects.* These are productivity gains that come from product and process improvement over time.

The emphasis on capex versus opex should vary by business segment. Upstream supply chain management should focus on a balance of cost containment and asset productivity, because of the relative importance of operating expense (opex) to capital expense (capex) (fig. I–3). Electric and gas utilities have the most capex relative to opex (3.7 dollars of assets for each dollar of operating expense),[2] so are biased toward asset productivity management (synchronization). Midstream and downstream activities have a greater proportion of opex per dollar of assets (1.3 dollars of assets for each dollar of opex), so supply chain management in midstream tends to focus more on operating cost reduction (rationalization) than on asset productivity (synchronization). Upstream companies are in between (2.3 dollars of assets per dollar of opex), so supply chain management in exploration and production (E&P) needs to achieve a finer balance between capex reduction and operating-cost containment.

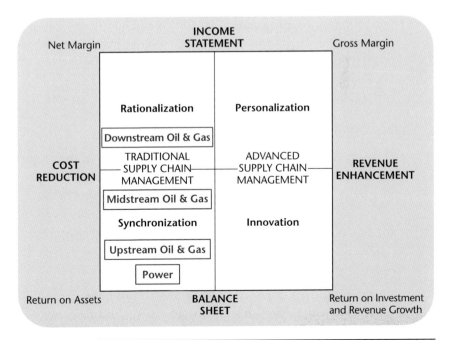

Fig. I–3. Prevalent supply chain strategy positioning of oil, gas, and power companies

Key supply chain processes

Typical cross-industry supply chain management processes include (fig. I–4): manufacturing and distribution site location (which in most industries is independent of the proximity to natural resources and is assumed to be via fungible modes of transport like truck, rather than specialized modes like very large crude carriers); demand planning; strategic sourcing; purchasing and inbound logistics; production; management and warehousing; and distribution operations.

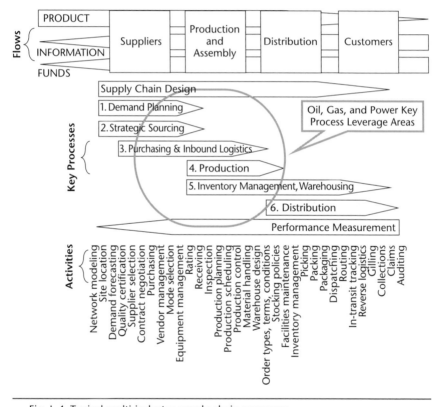

Fig. I–4. Typical multi-industry supply chain processes

However, the elements of supply chain strategy for oil and gas are more focused. Oil, gas, and power companies are asset intensive, so many of their key business processes have to do with asset procurement, installation, deployment, operation and maintenance. A cross-section of upstream and downstream oil and gas business processes and power plant management processes looks quite different from analogous processes at consumer goods companies, following the pattern outlined here:

- Negotiate or bid for and acquire land

- Acquire physical data (seismic, geologic, wind/wave)

- Model hydrocarbon resources/multiplant or power grid network potential

- Decide ownership structure (e.g., see chap. 1)

- Develop a project execution plan

- Design wells/processing routes/turbine systems

- Design facilities and installations

- Negotiate engineering, procurement, construction, and installation (EPCI) contracts

- Construct rigs/platforms/vessels/pipelines/refineries/ power plants

- Drill wells/build refineries

- Operate wells/refineries/power plants

- Maintain wells/refineries/power generation equipment

- Monitor and evaluate reservoir/asset performance

- Sell or turnover depleted assets or abandon/remediate unproductive assets[3]

Mapping the most relevant generic supply chain techniques[4] to the business processes that are the most relevant to oil, gas, and power supply chain success[3] yields a hierarchy of key supply chain processes. Distilling this integrated framework yields three major topic areas, which form the structure of each chapter in this book. The chapters are further subdivided into topics that apply uniquely to upstream, midstream, and downstream oil and gas and power. The following are three major subdivisions:

- Capex project supply chain risk mitigation
- Engineering and procurement of equipment and services at minimum cost and risk
- Operating-cost reduction

Performance management metrics and targets—asset and cost based

Ultimately, supply chain improvement initiatives result in higher net margin and return on assets (ROA) (fig. I–5). From a strategic point of view, the objective is to leverage effective supply chain management to outperform competitors as well as company historical performance on these dimensions. This creates shareholder value above expectations, which increases stock price (for publicly held companies).

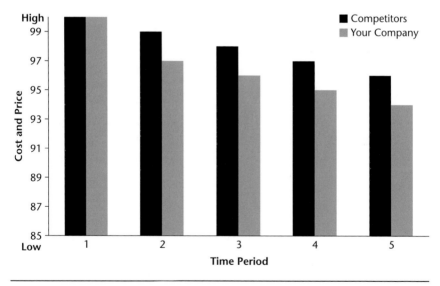

Fig. I–5. Supply chain cost and price reductions creating value above market

However, margin and asset performance improvements are typically measured at the project level, where they are sought in targeted areas such as the following:

- *Reduced upfront purchase cost.* Lower upfront cost is the most intuitive savings framework. However, tiered pricing, promotions, discounts, and volume rebates can make lowering upfront costs more complicated than it may seem.

- *Reduced operating cost.* Energy savings is a common way to reduce operating costs. Operating assumption variables can affect the savings, especially for large turbines and electrical distribution and control equipment.

- *Increased throughput or productivity.* Improvements that increase the speed of a process such as drilling or refinery expansions can lead to higher production overall. Savings for these types of improvements can be calculated on the basis of either cost savings or profit enhancement. Cost savings might be estimated by, for example, the reduction in the number of rig-days needed to drill a well. The savings per well then can be multiplied by the number of rigs in operation and the number of wells that need to be drilled over a period of time. If the increased productivity leads to reduced time to first production, then the benefit may be improved profitability. To quantify this benefit, take the number of extra days of production and calculated additional profit based on a typical well, then apply an average output price per unit to get the profit margin, and multiply the resulting benefit by the number of producing units.

- *Shorter lead times for the delivery of equipment or services.* For capital items, one might determine that order lead times are constraining production that would otherwise be occurring. In this case, the decrease in lead times can be multiplied by the weekly profit from greater production. For lower lead times on purchases that are held in inventory, inventory holding costs can be used as the basis for cost savings.

- *Increased reliability or uptime.* Longer mean time between failure means the same amount of production with less equipment, saving money by avoiding the need for equipment purchases. Benefits can be calculated according to how much less equipment is needed (in value, not in units).

Metrics should be consistent with the chosen supply chain strategy and the targeted benefits.

The combined effect of supply chain initiatives in practice raises corporate EVA at oil, gas, and power companies by 3.4%, from a benchmark of 8.4% to 11.8%. Operating cost management increases EVA to 10.7%, and if capacity is constrained, the benefit rises to 11.8% by unlocking additional capacity with the same asset base. Greenfield projects can realize much higher savings owing to the ability to design in savings early (fig. I–6). Benefits typically include, in order of their impact on EVA,[5]

- 13% reduction in capital costs. This would increase total corporate EVA by 0.8%, from 8.4% to 9.2%.

- 9.8% increase in total annual sales revenue. This is due to debottlenecking and increased throughput capacity and would increase total corporate EVA by 0.5%, from 8.4% to 8.9%. If capacity is constrained and all improvements resulted in output that could be sold immediately, total sales would increase by 45%, and EVA would increase by 1.9%.

- 1.0% reduction in total operating cost. This would increase total corporate EVA by 0.6%, from 8.4% to 9.0%.

- 1.6% reduction in inventory holding cost. This would increase corporate EVA by only 0.01%, from 8.4% to 8.41%, owing to the asset-intensive nature of the business.

Some improvement initiatives can be deployed at the expense of added risk, so risk must be measured and managed in conjunction with supply chain improvement programs. There are many complicated ways of measuring economic risk, most involving statistics and probability (and there are many more ways to measure operational risk; those will be covered in the next section). In general, executive managers discount probabilistic financial analyses, preferring to trust their gut feeling around scenarios, of which one usually represents the most likely, or *baseline*, scenario. A *tornado diagram* is a relatively simple and intuitive way to measure and display the economic risk impact related to supply chain programs (fig. I–6).[6]

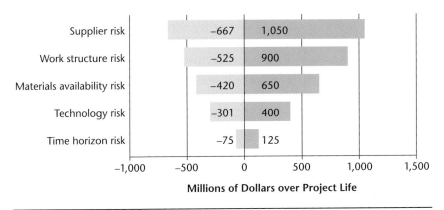

Fig. I–6. Quantification of supply chain management risks

Examples of supply chain savings/success

Each year Boston Strategies International (BSI) conducts a not-for-profit, in-depth global supply chain benchmark study in which diverse companies respond from countries as widespread as Australia, Germany, India, Italy, Nigeria, Norway, South Africa, and the United States. Since its inception, over 500 companies have applied for the award, given in each of three categories: oil company, equipment provider, and service provider.

Qatar Fuel, Chevron, Baker Hughes, and Bharat Petroleum are some of the previous years' winners. Qatar Fuel won the 2008 award for overall supply chain and operations excellence by leading industry performance in 15 different supply chain metrics—for example,

- 58% return on capital employed
- 46% RONA
- Six Sigma order and delivery cycle-time reliability
- >99% orders delivered by time customer requests
- >99% of orders delivered by the time committed to
- >99% stock accuracy

- >97.6% uptime
- >95% first-pass yield
- 5.6% cost of order fulfillment as percentage of order value

Chevron (USA) won in 2010 for its ability to extend its internal lean Six Sigma success to its suppliers. Its lean Six Sigma program, which officially started as a grass roots initiative in 2000, created financial benefit of $250 million for Chevron in 2008, $400 million in 2009, and an estimated $500 million in 2010. Chevron's first pilot project in 1999 reduced operating costs at a California water treatment plant by 30%. Hundreds of projects across the United States, United Kingdom, Angola, and Indonesia contributed to the achievement. An internal consulting group helps Chevron business units start and individual programs mature. As the programs mature, Chevron actively engages its suppliers and contractors, both to further existing improvement projects and to develop their own continuous improvement programs.

Also in 2010, Baker Hughes won in the equipment provider category, based on first-year achievements under its supply chain transformation strategy, with cost reduction targets of $100 million per year for 2010, 2011, and 2012. It plans to achieve these savings through supplier categorization and rationalization, strategic sourcing to shorten the supply chain, and lean manufacturing improvements. The firm created a global supply chain and manufacturing organization in April 2009, with Arthur Soucy, formerly Pratt & Whitney's vice president of Global Supply Chain, in charge.[7] Baker Hughes also committed to $75 million in annual savings for 2011 and 2012 from its integration of BJ Services' operations, which began in September 2010.[8]

Bharat Petroleum won the 2009 award in the oil company category, for its superior order accuracy, inventory management, and overall asset productivity. Other examples of supply chain cost management include Nalco and BASF. Nalco reduced its costs by $122 million in 2010 by using Six Sigma; the company has a team of Six Sigma blue belts tasked with cutting operating costs by $100 million per year.[9] BASF saved $300 million in 2010—half of the $600 million it has saved to date from the program called *NEXT*, instituted in 2000, that aims to save a total of €1 billion.[10]

Organizing Supply Chain Management Activities

Discrete manufacturing firms such as consumer products companies typically have senior executives with titles such as chief supply chain officer, chief procurement officer, or chief logistics officer. Their staff includes such departments as procurement, production, order fulfillment, materials management, demand planning, logistics, customer service, production control, and transportation.[11] This is also the case with many suppliers to the oil and gas industry. Some oil, gas, and power operators bring on staff from other industries. Baker Hughes, as mentioned above, hired a chief supply chain officer from Pratt & Whitney; others have hired chiefs with automotive backgrounds, and many have corporate departments like those mentioned above.

Utilities often have similar organizational structures, with prominent groups called procurement, fleet/transportation, and materials/inventory management. However, many oil and gas organizations assign most operational supply chain responsibilities to staff with formal titles like engineering, health and safety, drilling, and other departments that might not be recognizable as containing supply chain management content. For example, supply chain management activities—such as project risk mitigation, standardization and simplification of specifications, minimization of capex procurement, supply risk mitigation, constraints management and throughput improvement—are not apparent or recognizable when they occur within upstream oil and gas departments with names such as project development, construction management, commissioning, project services, drilling, surface operations, and planning; and typical downstream oil and gas departments such as export/transportation (or "supply"), plant engineering, and marketing.[12]

Organizational fragmentation and high capital expenditure reinforce the need for continuous learning and training of operations staff in supply chain principles and practices. Partly as a result of the organizational fragmentation, the expenditure per supply management employee is lower in the electric power industry than in a cross-section of other industries (table I–2). In contrast, because of the preponderance of large capital projects, the expenditure per supply management employee is higher in oil and gas than in other industries. Both fragmentation and high expenditure are good reasons for continuous learning and training of operations staff in supply chain principles and practices. Some organizations foster supply chain management training and staff development. Saudi Aramco holds an annual Supply Chain Symposium in conjunction with King Fahd University of Petroleum and Minerals, to bring together key stakeholders and decision-makers from suppliers, academia, and professional associations. By assembling local and world-renowned experts and exhibitors in the fields of supply chain and risk management, the initiative aims to develop a common language about supply chain management that can support extended supply chain management.[13]

Table I–2. Managed spending per supply management employee (*Source:* CAPS Research 2011. Reprinted with permission from the publishers, Institute for Supply Management™ and W. P. Carey School of Business at Arizona State University.)

	Oil & Gas	Electric Power	All Industries
Mean	17.9	10.7	
Minimum	2.4	2.0	
Maximum	28.1	33.6	
Median	19.5	8.3	15.6

Note: Data are in millions of U.S. dollars.

First Principles for Supply Chain Design and Improvement

A series of first principles can help to define the perimeter of potential solutions to the array of problems that are addressed in the capex and the opex sections. The following are three such overriding principles:

- Supply chain design should consider the end-to-end supply chain.
- Supply chain solutions should minimize life-cycle cost by balancing capital and operating costs.

- The structure of partner relationships should minimize contract, operating, and environmental risk.

In supply chain design, alignment with suppliers of goods and services and with customers increases the likelihood that targeted benefits will be achieved. Suppliers' collaboration and adherence to a common and agreed set of practices and principles is essential to achieving cost reduction and risk minimization goals; conversely, divergence from such goals increases the chances of mishaps, as evidenced through BP's failed coordination with Transocean, Halliburton, and Cameron, which contributed to the Macondo spill. In addition, coordination of supply-and-demand forecasts and production and inventory levels can dampen the *bullwhip effect*, an oscillation and amplification of demand throughout the value chain that adds 10% to the ultimate price of products in the oil and gas supply chain; as demand increases, oil price rises, which causes producers to expand capacity, forcing prices down and depressing demand for equipment, in a feedback loop.[14] Oil and gas drilling activity fluctuates about three times as much as production of refined product, indicating (among other things) that if refiners coordinated more with E&P, then there would be less volatility up and down the supply chain.[15] The bullwhip effect can be further quantified in the upstream oil and gas supply chain, finding that it increases the cost of gasoline by 10% as oil producers, oil refiners, heavy equipment suppliers, and their component suppliers pass on the costs of inventory overages and shortages, poorly timed capacity investments, and inflationary prices (see app. A). A similar problem occurs in a multi-tier electrical distribution system, where peak demand requires the use of *peaker* generating plants at the upstream end of the supply chain, which can be activated to cope with fluctuations in demand. Supply chain planners and operations researchers have designed hundreds of models to balance and optimize fluctuating demand through a system, and this book will show some of their applications (fig. I–7).

Cost of Supply Chain Misalignment

Refining

Total cost for a refiner

- Cost of utilizing capital investment 1
  - Interest rate
  - <Capacity utliization rate 1>
  - <Oil price>
  - <Potential production 1>

- Opportunity cost of lost production 1
  - Profit margin 1
  - <Unfulfilled demand 1>
  - <Oil price>

- Cost of holding inventory
  - <Oil price>
  - <Inventory 1>

- Excess price paid to purchase equipment 1
  - Addition of new equipment 1
  - <Equipment price>
  - <Stable long-term equipment price>
  - <Output Input ratio 1>
  - <Capacity adjustment 1>

Oil & Gas Production (Oil Rigs)

Total cost for oil & gas producers

- Cost of utilizing capital investment 2
  - Interest rate
  - <Capacity utliization rate 2>
  - <Price of crude oil from rigs>
  - <Potential production 2>

- Opportunity cost of lost production 2
  - Profit margin 2
  - <Unfulfilled demand 2>
  - <Price of crude oil from rigs>

- Excess price paid to purchase equipment 2
  - Addition of new turbines 2
  - <Turbine price>
  - <Stable long-term turbine price>
  - <Output Input ratio 2>
  - <Capacity adjustment 2>

Hot Section & Blade Manufacturing (Component Suppliers)

Total cost for a hot section manufacturer

- Cost of utilizing capital investment 4
  - Interest rate
  - <Capacity utlization rate 4>
  - <Hot section price>
  - <Potential production 4>

- Opportunity cost of lost production 4
  - Profit margin 4
  - <Unfulfilled demand 4>
  - <Hot section price>

- Excess price paid to purchase equipment 4
  - <Excess price paid to purchase equip 3>

Turbine Manufacturing (OEMs)

Total cost for a turbine manufacturer

- Cost of utilizing capital investment 3
  - <Interest rate>
  - <Capacity utliization rate 3>
  - <Turbine price>
  - <Potential production 3>

- Opportunity cost of lost production 3
  - Profit margin 3
  - <Unfulfilled demand 3>
  - <Turbine price>

- Excess price paid to purchase equipment 3
  - Addition of new hot section 3
  - <Hot section price>
  - <Stable long-term hot section price>
  - <Output Input ratio 3>
  - <Capacity adjustment 3>

Fig. I–7. System dynamics model of a four-tier upstream oil and gas supply chain (*Source:* Jacoby, David. 2010. The oil price bullwhip: problem, cost, response. *Oil & Gas Journal*. March 22: 20–25.)

Supply chain choices must minimize life-cycle cost by balancing capital and operating costs over the relevant time horizon of the investment. Many decisions in these long investment horizons involve trade-offs between initial capex and ongoing opex. It is easy to save money on initial capex if one is willing to pay more in opex, and many suppliers would prefer to sell solutions that result in perpetual income for their companies. Although the idea is simple, execution of true life-cycle cost analysis is difficult for two reasons. First, with large and complex projects (see fig. I–8), the life-cycle cost components are often so extensive that it is frequently hard to decide which ones are relevant. Some may appear to be unrelated (e.g., an investment in an extra module of a system of enterprise resource planning that needs to be in place to communicate with key suppliers), interrelated or contingent (investments may need to take place only if a milestone is reached), or sunk costs (e.g., dredging to lay a cable, where the dredging may have had to be done anyway). The second challenge is getting reliable enough data to make good trade-offs. Frequently the data are simply unavailable (e.g., reliability performance of a new technology platform for a gas turbine or life span of wind turbine blades made in China versus comparable blades made in Europe).

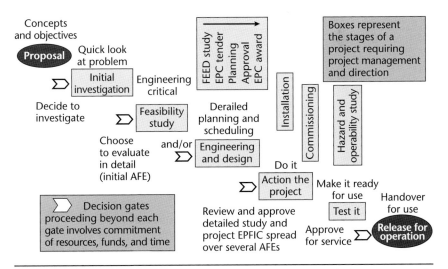

Fig. I–8. Typical oil and gas production project framework divided into stages and gates (*Source*: Mokhatab, Poe, and Speight, 2006, p. 548.)

Third, supply chain management should minimize contract, operating, and environmental risk. Because of their long time frames, oil, gas, and power plant projects are inherently risky. A typical upstream or midstream project time frame for an LNG export facility could be seven years (see fig. I–9).[16] Longer time frames bring more market risk, because costs and prices will change over the time horizon of construction, installation, and operation. The magnitude of investment typically brings supplier risk. The milestones bring *go/no-go* breaking points, which could halt the project, increasing or reducing the rate of return relative to initial forecasts. A particular challenge arises when suppliers require financial commitment before the final investment decision (FID), to deliver the product or service by the deadline. The role of supply chain management should be to minimize these risks. Risk minimization may be at odds with management's leanings, especially if rapid implementation would accelerate substantial rewards or if savings could be achieved either by reducing the investment of time or money or by discounting negative environmental externalities.

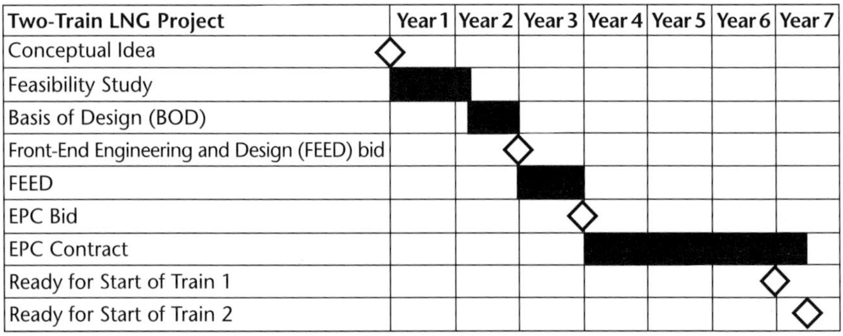

Two-Train LNG Project	Year 1	Year 2	Year 3	Year 4	Year 5	Year 6	Year 7
Conceptual Idea	◇						
Feasibility Study	■						
Basis of Design (BOD)		■					
Front-End Engineering and Design (FEED) bid		◇					
FEED			■				
EPC Bid			◇				
EPC Contract				■	■	■	
Ready for Start of Train 1						◇	
Ready for Start of Train 2							◇

Fig. I–9. Typical project time frame for an LNG export facility (*Source:* Mokhatab, Poe, and Speight, 2006, p. 548.).

Notes

1 Mokhatab, Saeid, and William A. Poe, and James G. Speight. 2006. *Handbook of Natural Gas Transmission and Processing*. Burlington, MA: Elsevier. P. 8.

2 Boston Strategies International analysis of data from Thomson Financial, on a "snapshot" basis using latest available 12 months of data for each company.

3 Adapted from Raymond, Martin S., and William L. Leffler. 2005. *Oil & Gas Production in Nontechnical Language*. Tulsa, OK: PennWell.

4 For a generic view of supply chain processes, see The Economist Guide to Supply Chain Management. The Guide specifies four supply chain strategies and 22 methods, of which the following five are of particular relevance for oil, gas, and power companies: (1) strategic sourcing and outsourcing using integration, scale and value engineering techniques; (2) lean manufacturing, in so far as it is focused on waste reduction and quality management; (3) standardization and simplification of specifications; (4) constraints management and throughput analysis; and (5) risk mitigation.

5 EVA measures profit after tax less the true cost of capital employed. It is calculated as follows: NOPAT − (capital × cost of capital), where NOPAT is the net operating profit after taxes.

6 Raymond and Leffler, 2005.

7 Boston Strategies International. Baker Hughes wins Boston Strategies International's 2010 Oil and Gas Award for Excellence in Supply Chain Management. http://www.bostonstrategy.com/images/Baker_Hughes_Wins_2010_BSI_Supply_Chain_Award.pdf (accessed September 22, 2010).

8 FD (Fair Disclosure) Wire. 2010. Q3 2010 Baker Hughes Incorporated earnings conference call—final FD. Wire. November 1. P. 6.

9 Thomson Reuters. NLC − Q4 2010 Nalco Holding Company Earnings Conference Call. Thomson Reuters. http://www.google.com/url?sa=t&rct= j&q=&esrc=s&source=web&cd=2&ved=0CG8QFjAB&url=http%3A%2F%2 Fphx.corporate-ir.net%2FExternal.File%3Fitem%3DUGFyZW50SUQ9MzY2 OTM1NHxDaGlsZElEPTQxMTk4MHxUeXBlPTI%3D%26t%3D1&ei=i32-T-vcJIrS2QX6x8iWDw&usg=AFQjCNGH7PSmPHYuqvSdxbGsWfke6uEZWg &sig2=EMIzsYeA6ZN_U3eeilS7jg (accessed February 9, 2012).

10 Basf. 2011. Sustainable improvement of cost base. Basf Corporate Web site. http://www.basf.com/group/corporate/en/investor-relations/strategy/cost-of-capital/cost-reduction (accessed May 19, 2012).

11 Jacoby, David. 2009. *Guide to Supply Chain Management*. New York: Bloomberg. P. 161.

12 Graham, Mark, Mark Cook, and Frank Jahn. 2001. Project and contract management. *Hydrocarbon Exploration and Production*. 55. P. 329; Paik, Jeom Kee, and Anil Kumar Thayamballi. 2007. *Ship-Shaped Offshore Installations: Design, Building, and Operation*. New York: Cambridge University Press. P. 38.

13 Saudi Aramco. 2010. Managing the supply chain: A complex but vital task. http://www.saudiaramco.com/en/home/news/latest-news/2010/managing-the-supply-chain--a-complex-but-vital-task.html#news%257C%252Fen%252 Fhome%252Fnews%252Flatest-news%252F2010%252Fmanaging-the-supply-chain--a-complex-but-vital-task.baseajax.html (accessed January 14, 2012).

14 Mashayekhi, Ali. 2001. Dynamics of oil price in the world market. Paper presented to the 19th International Conference of the System Dynamics Society, July 23–27, Atlanta, GA. http://www.systemdynamics.org/conferences/2001/papers/ (accessed September 10, 2010).

15 Sterman, John. 2006. Operational and behavioral causes of supply chain instability. *The In The Bullwhip Effect in Supply Chains*. Ed. Octavio A. Carranzo Torres and Felipe A. Villegas Morán. New York: Palgrave Macmillan.

16 Tusiani, Michael D., and Gordon Shearer. 2007. *LNG: A Nontechnical Guide*. Tulsa, OK: PennWell. P.125.

Part 1

1

CAPEX PROJECT SUPPLY CHAIN RISK MITIGATION: PRINCIPLES AND METHODS

Capital projects involve many risks that are typically absorbed and thus are considered to be part of doing business. Some risks are inevitably left unquantified, and some projects proceed despite too much or too little risk (over- or underinvestment). Shareholders sometimes absorb the risks without knowing about them.

Types of Choices Involving Risk

Capital projects involve market (price and volume) risk, materials supply risk, supplier risk, construction risk (sometimes off-loaded to an engineering, procurement, and construction [EPC] firm), and operational, supplier, technology, political, and regulatory risks.[1] Capital project managers can seek market risk analyses from a number of specialist consulting firms; political and regulatory risk are special types of risk that extend well beyond supply chain management and therefore will not be covered in this book.

Of predominant interest to supply chain managers are the size and duration of exposure to supply commitments (time horizon risk), technology risk, materials availability risk, supplier risk, contract risk (financial exposure), and operational risk.

Time horizon risk

Time horizon risk quantifies exposure on commitment over more or less time. By choosing a time horizon for project investment, project managers are implicitly embedding a supply chain risk. If the time frame is too short, suppliers may not reap enough economies of scale or learning-curve benefits to be able to lower costs enough to meet targets. If it is too long, the owner or operator may effectively be giving the benefits of productivity gains, learning effects, and technology innovations to suppliers. How can supply chain managers assess time horizon risks, and what can they do about them?

Technology risk

Because of the complexity and incertitude in quantifying technology risks, many companies prefer to use existing technologies with well-established risk profiles. However, making large, long-term capital investments on the basis of stale technology is all but assuring lower-than-market rates of return potentially for decades. Is there a methodology for quantifying this risk?

Materials and labor availability risk

Embedded assumptions about materials availability, service prices, and skilled trade wages have caused many cases of force majeure. In one case, a manufacturer of expandable tubulars experienced a bottleneck in outsourced machining capacity, which delayed production and incited a bidding war for machine time, which raised costs. Is there any way to mitigate potential or actual shortage situations?

Work structure risk

When working with large projects involving many systems and subsystems, operators can choose from a spectrum of solution-bundling choices. They can opt to design, build, and operate in-house and can even unbundle the whole supply chain down to the component level. Alternatively, they can hire an EPC firm to engineer, procure, construct, install, and operate, but would effectively own none of the value chain activities. Unbundling may save the operator money by avoiding integrator markups, but this internalizes the risks of nonperformance and liabilities for faulty work, both of which can be so large in some cases as to have a substantial impact on the operator's financial health and even viability. Which is the right balance of risk and cost-effectiveness?

Supplier risk

Supplier failure can sometimes have a disastrous effect on a project; consequently, many companies have extensive prequalification processes and require performance bonds. Supplier risk can also apply in other situations outside of outright insolvency. The choice of companies on the bid slate has a large impact on the competitiveness of bids, and the structure and intimacy of the ensuing relationship has a large impact on the cost efficiencies and performance levels achieved during the project.

General Approaches to Managing Supply Chain Risk

Supply chain policies, processes, systems, and organizational structures can be used to develop a strategic approach to risk (fig. 1–1).

The easiest and in many cases the most effective risk management strategy is to *avoid* risk entirely by passing it through to customers. BSI studied how manufacturing companies dealt with the rising cost of energy, using eight risk-mitigation strategies.[2] The study found (not unsurprisingly) that when customers allowed companies to institute price increases commensurate with the increased costs, this largely mitigated the actual cost increases over the period studied.

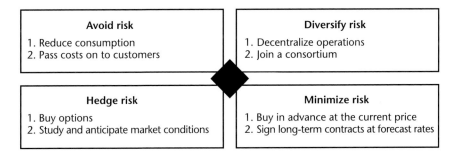

Avoid risk
1. Reduce consumption
2. Pass costs on to customers

Diversify risk
1. Decentralize operations
2. Join a consortium

Hedge risk
1. Buy options
2. Study and anticipate market conditions

Minimize risk
1. Buy in advance at the current price
2. Sign long-term contracts at forecast rates

Fig. 1–1. Strategies for managing supply price and availability risk (*Source:* Jacoby, David. Supply Risk Management." Webcast. November 19, 2007. Page 9.)

The study also found that there were two ways to pass through the price increases—through surcharges or by embedding them in the base sales price. Adding an energy surcharge resulted in a cost-neutral position. Cost increases were passed through when they occurred and were retracted when they went away. An example of this would be the pricing formula for LNG in Europe, where increases in exploration and production costs are passed through by linking the LNG price to changes in the prices of fuels, such as gas oil, low-sulfur fuel oil, and coal, and adjusting the price of LNG to match the change in the price of those fuels.[3] The other approach was to embed the cost increase in the base price. This proved to create a lucrative profit center, a strategy that was acknowledged by many interviewees. It not only shielded buyers from price increases but also resulted in windfall gains because energy prices subsequently declined while those increases remained structurally embedded in the sales price.

Another approach to risk avoidance is to off-load it to project managers (e.g., EPC contractors) or use lump sum/turnkey (LSTK) contracting parameters that shift any risk to the suppliers. EPC contracts typically delegate responsibility to a contractor for making sure the facility functions as intended. If for any reason the facility does not work as planned, then the EPC, rather than the operator, is responsible for remedying the situation. This assures the buyer of a single point of contact, of a guaranteed result, and if the contract is for a fixed price, of no cost risk.

Under an LSTK contract (which can also be with an EPC firm), the *lump sum* aspect makes any overage the responsibility of the contractor. The *turnkey* aspect underlines the condition that the contractor must deliver a working facility, regardless of cost-overrun status. About 6% of total petroleum industry expenditure runs through EPC contractors, according to Institute for Supply Management (ISM) benchmarks.[3] The downside of most EPC and LSTK contracts is that they often embed either cost adders or buffers that protect the contractor against contingencies, and these buffers can exceed the average cost of the unexpected events. Also, EPC contractors may have an incentive to overengineer, since almost 60% of their contracts are on a time-and-materials basis.[4] For example, one operator complained that EPC firms on a time-and-materials contract to build a water treatment plant would have an incentive to recommend an expensive bottom-mounted aeration unit, instead of a cheaper and more efficient top-mounted system, because it would boost their revenue.

A second strategy for managing risk is to *diversify* or *share* it, typically by sharing it between the buyer and the supplier. Buyers prefer unlimited liability for the supplier while it is working on the buyer's site, whereas suppliers usually negotiate for liability caps at contract value or a percentage thereof. Negotiation of the liability caps can shift risk between the two parties. For example, on a $2 billion construction contract, the negotiated cap might end up being 20% of the value of the contract.

A third strategy is to *minimize* risk. Long-term agreements often minimize risk for both parties. A recent study showed that buyers realize—and suppliers concede—3%–8% savings, on average, by negotiating long-term agreements (fig. 1–2).[5] This price reduction reflects the volatility that was removed from the supplier's future sales stream. However, depending on how it is written, a long-term contract can also increase risk, by locking buyers or suppliers into bearing the cost of underlying materials price increases. The more stable long-term contracts share or index these costs, to ensure that one party will not suffer and eventually have to default under the weight of a cost increase.

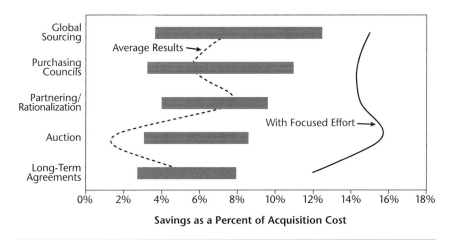

Fig. 1–2. Savings from long-term agreements (*Source:* Jacoby, David. 2005. *High-Impact Sourcing: Precision-Guided Strategies for Maximum Results.* White Paper. Boston Strategies. http://www.bostonstrategies.com/images/BSI_-_SS2_High-Impact_Sourcing.pdf.)

The fourth strategy is to *hedge* risk. Contract price indexation can accomplish this. For example, a $1 billion specialty minerals processor indexes 10-year agreements to its costs, effectively negating any cost inflation and still allowing it to pass through an escalation of costs each year on top of that risk-hedged rate. Other companies buy products that contain extensive amounts of steel by specifying the tons of steel they contain; this determines the price per ton at the time of the order, which price then increases or decreases on delivery to the customer on the basis of the change in steel price over the order delivery lead time. Battery suppliers do the same thing with lead costs, and cable and some motor suppliers do this with copper prices. Polyvinyl chloride pipe is sometimes indexed to resin prices. Other hedging techniques leverage financial options (traded on a mercantile exchange), futures contracts (traded on some stock exchanges), forward contracts (between two parties), and real options (a plan or contract that allows investments to be triggered, changed, or cancelled at certain milestone points).

A hybrid strategy is to *anticipate* risk. Some firms track and/or forecast the market, a practice known as *supply market intelligence*, in an attempt to select the most effective risk-mitigation strategy for each circumstance. This is analogous to an intelligent demand-forecasting module that evaluates a variety of forecasting methods each period and applies the one that would have had the least forecast error over the past several periods. World-class supply market intelligence includes seven dimensions of analysis (fig. 1–3):

- Capacity and scale
- Product and process technology
- Global opportunities and risks
- Costs
- Demand
- Channels and supply chain
- Competitive dynamics

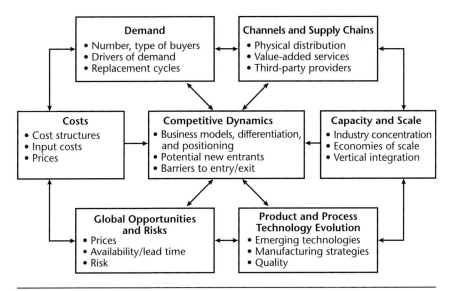

Fig. 1–3. Framework for supply chain market intelligence (*Sources*: Jacoby, David. "Focus on the Right Suppliers," *Purchasing*. February 27, 2005. P. 64; Jacoby, David. "Swagging the Biggest Decisions You Make," *Purchasing*. May 5, 2005. P. 60; Jacoby, David. "Measuring Sourcing Performance: What's the Mystery?" *Purchasing*. June 2 2005. P. 60; Jacoby, David. "Logistics of Global Outsourcing," presentation to APICS Providence chapter. Oct. 19, 2005. Pp. 29–36.)

Determining the Optimal Term of Commitment

The contract term has as much impact on the total cost and risk profile of the project as the choice of whether to insource or to outsource, although it is often decided by gut feel. A priori, it is intuitive that commitments should be long term because many of the projects' basic parameters are long term—namely,

- *Long time to plan.* Projects that take a long time to plan often require long forward views and extensive cost and economic estimations.

- *Long time to build.* If investments take a long time to build, then they require a long planning horizon. The longer the time to build is, the greater the risks are that important planning assumptions will be wrong or will become wrong during the building time horizon.

- *Deferred revenue streams.* Along with large capital expenditures usually come deferred revenue streams and unpredictable future cost levels.

- *High exit costs.* To the extent that industries or projects are costly to exit, the planning horizon extends even further.

In fact, more than half of the oil and gas companies surveyed in a 2009 study said that a long-term stable pricing environment would help them establish steadier prices and operating profits, thereby allowing them to minimize layoffs during downturns and rehiring during upturns—and consequently reducing long-term operating costs. A third of the companies surveyed said that long-term stability would allow them to make more predictable research and development (R&D) investments, which would result in higher exploration, refining, and distribution productivity owing to faster and more consistent advances in oil and gas equipment technology.[6]

The question remains, How do you know which term is right— 10, 20, 30, or 40+ years? The term of commitment is sometimes decided arbitrarily. Oil and gas, power, water, and transport infrastructure providers frequently sign 10-year contracts, 20-year operating agreements, and 40- or 50-year development and operating contracts. The term should be driven by the expected trend and volatility in the underlying cost and rate structures and by the economic dynamics that can be expected or forecast to change over the time frame of the agreement (economies of scale, scope, standardization, learning-curve benefits, technology advances, and competition), as illustrated later (see fig. 1–7).

One approach to optimizing contract term is to use an optimization model (fig. 1–4). Boston Strategies International has developed a contract term model that uses the historical volatility in price movements and the likely savings that could be obtained from a long-term contract to test the expected net benefits at a spectrum of possible intervals on a long time horizon. The model uses previous engagement history to define probability distributions for the volatility and the savings and generates an optimal contract term through Monte Carlo simulations.

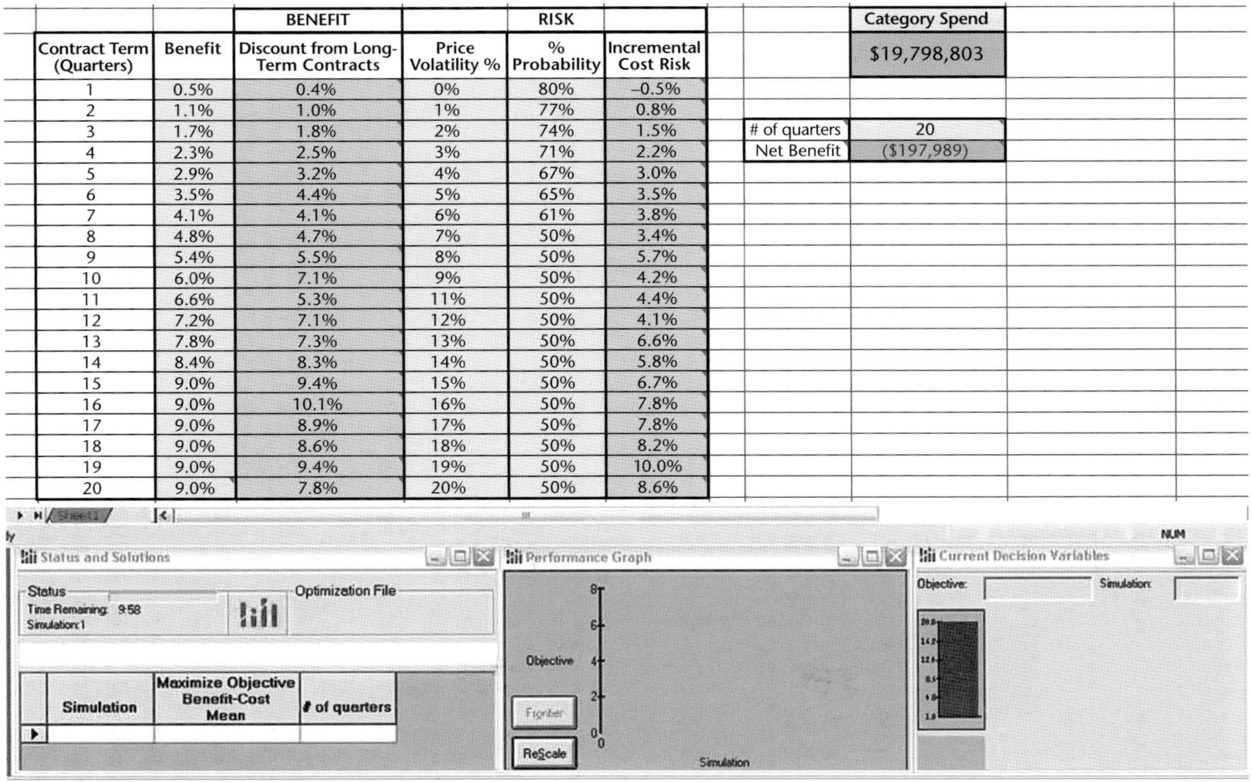

Contract Term (Quarters)	Benefit	BENEFIT Discount from Long-Term Contracts	Price Volatility %	RISK % Probability	Incremental Cost Risk	Category Spend
						$19,798,803
1	0.5%	0.4%	0%	80%	-0.5%	
2	1.1%	1.0%	1%	77%	0.8%	# of quarters 20
3	1.7%	1.8%	2%	74%	1.5%	Net Benefit ($197,989)
4	2.3%	2.5%	3%	71%	2.2%	
5	2.9%	3.2%	4%	67%	3.0%	
6	3.5%	4.4%	5%	65%	3.5%	
7	4.1%	4.1%	6%	61%	3.8%	
8	4.8%	4.7%	7%	50%	3.4%	
9	5.4%	5.5%	8%	50%	5.7%	
10	6.0%	7.1%	9%	50%	4.2%	
11	6.6%	5.3%	11%	50%	4.4%	
12	7.2%	7.1%	12%	50%	4.1%	
13	7.8%	7.3%	13%	50%	6.6%	
14	8.4%	8.3%	14%	50%	5.8%	
15	9.0%	9.4%	15%	50%	6.7%	
16	9.0%	10.1%	16%	50%	7.8%	
17	9.0%	8.9%	17%	50%	7.8%	
18	9.0%	8.6%	18%	50%	8.2%	
19	9.0%	9.4%	19%	50%	10.0%	
20	9.0%	7.8%	20%	50%	8.6%	

Fig. 1–4. Optimization model output for the optimal contract term (illustrative) (*Source:* Jacoby, David. 2007. "Budgeting for Volatility." Presentation to Demand Planning and Forecasting Best Practices Conference, April 27, New Orleans. Institute of Business Forecasting. P. 32.)

In several cases, the simulation yielded optimization results of more than 20 years. One study showed that it takes 22 years for an initial price shock to work its way through a capacity-price-cost-sales cycle, implying that long-term agreements in this context should be at least this long to completely eliminate market risk.[7]

An alternative is to establish an initial agreement that includes structured options for expansion, contraction, modification, and cancellation as the project unfolds. This approach has the advantage of yielding more informed decisions, by relying less on forecasts or bets. It also is economically more efficient and, on average, results in higher net present values (NPVs), since it lowers the cost of capital investment by maintaining the upside opportunity while limiting the downside risk.

There are four relevant types of options, each of which corresponds to a type of analysis in real-options theory:

- Option to expand, called a *call option*
- Option to contract, called a *put option*
- Option to terminate, called an *abandonment option*
- Option to change the parameters, called a *switching option*

Option to expand

In the context of supply chain management and procurement contracts, options to expand are often called *capacity reservation agreements*. An example is an option to place or increase the number of orders or the size of orders, thus increasing the supplier's share of the expenditure. Suppliers will often grant a guaranteed capacity reservation on a subsequent order once an initial order has been confirmed. The contract typically specifies a fixed quantity of firm orders and an option to obtain additional units at a future time for an agreed price. Once an agreement is signed, suppliers are often willing to be flexible regarding the date when the option may be *called*, so long as it has adequate capacity; however, there may be constraints on the specifications and ancillary terms (financing, insurance, warranty, etc.) that the supplier can accommodate, depending on production capacity and configuration at the time the call option is exercised.

The value of options agreements are rarely publicly disclosed, since they reveal executives' expectations about their companies' future expected spare capacity, which could affect pricing on other sales. Most deals seem to be perceived to clear at 5%–10% of the normal purchase cost. Baker Hughes has optioned the rights to any barite found at Bravo Venture Group's new mine in central Nevada.[8] This also happens with aircraft, installation vessels, turbines, and a variety of other types of engineered equipment. (For additional examples, see the "Project Risk Mitigation" section chap. 8.)

Option to contract

The analogue to an option to expand is an option to reduce commitment. While these options can also be negotiated and their value can be quantified using real-options methods, flexible capacity strategies on the part of the supplier can also increase downward flexibility. At the project-planning stage, one way to make capacity flexible is to outsource peak load to a third party. This helps to keep capacity utilization steady and increases the reliability of service levels during both peaks and troughs in demand.[9]

Option to terminate

Options to terminate can also be built into programs and contracts. However, rather than valuing and articulating the option to terminate, in some cases, a termination-for-convenience clause may be used to provide the same benefit. For example, one oil company's LSTK contract stipulates that "the Company may, at any time, terminate any part of the Works or all remaining Works without giving any reason therefore, by giving written notice to the Contractor specifying the part of the Works to be terminated and the effective date of termination if different from the date of said written notice."[10]

Option to change

Similarly, almost all contracts have options to change orders at the discretion of the buyer. A change may be supplying a different product than the one originally ordered. Other frequently experienced changes include changes to design or specifications, changes to delivery timing, alteration of subcontracting rules or approval procedures, and assignment of the contract.

Technology Choices and Options

Technology decisions are a perennial challenge in oil, gas, and power generation. While new technology generally brings higher output, it almost always comes with a higher price tag and a risk that it will not deliver the anticipated benefits or that it will have new product complications, such as recalls or component failures. Since the investments are so large, it is hard to establish a verifiable pilot program that demonstrates the feasibility of the new technology at the scale at which it will actually be deployed. Furthermore, since the lead times are so long, suppliers cannot always invest in the extensive research and development needed to get to the next generation without prelaunch financial commitments from operators. Hence, a chicken-and-egg situation prevails, dampening innovation and putting a drag on productivity improvements.

Regardless whether it relates to subsea compressors, tension-leg platforms, tripod foundations, or next-generation turbines, a proper evaluation of the technology should take into account the following:

- Current and projected acquisition cost, including economies of scale and learning-curve effects
- Improvements in effective performance (e.g., power ratings) over time
- Operating, repair, and maintenance costs
- Infant mortality rates
- Supplier viability risk factors

Based on the total rolled-up cost over the time horizon of the project and the risk factors for the most strategic suppliers, project management can decide either to negotiate the desired outcome with target suppliers or, if it is an option, to frequently re-source newer technologies as the market produces them.

Learning-curve effects drive down the cost per unit of output for new technologies along a sometimes predictable curve (fig. 1–5). For example, wind turbines have brought dramatically higher power and lower costs to such an extent that, using in a scientific methodology reminiscent of Metcalfe's law, the time period has been calculated over which the cost can be expected to be cut in half. In the wind power industry, this is called the *progress ratio*.

Operating, repair, and maintenance costs typically decline as technology platforms and as new models mature (fig. 1–6). The initial period after model introduction has a higher repair and maintenance cost owing to infant mortality. Statistically, the risk of unplanned failure decreases as the number of units produced and put into operation increases, because of learning in the production process, actual performance, and customer feedback. Sometimes the decrease in costs can be considerable. For a long time, the cost per flight hour of Pratt & Whitney JT8 jet engines declined as the number of years since their launch increased, and GE's jet engines exhibited a similar phenomenon.

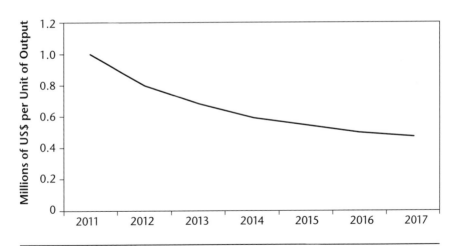

Fig. 1–5. Illustrative reduction in cost per unit over time

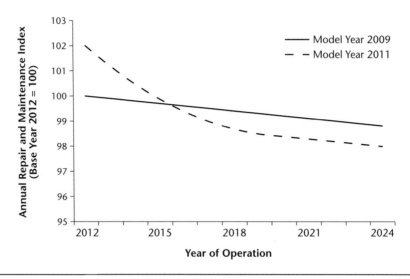

Fig. 1–6. Repair and maintenance costs as a function of new-model-introduction year (illustrative)

If the slope of the total cost curve (including economies of scale and all the other factors mentioned above) is projected to be downward and steep, it may be hard to negotiate such a large decrease with suppliers. In this case, operators need to decide whether to bring facts to the negotiating table and hammer out a long-term agreement based on a sharp projected decrease in long-term costs or whether re-sourcing on a periodic basis would be a more effective way of achieving the long-term cost target.

All of this generally supports the conservatism of many oil, gas, and power companies when it comes to new technology. Importantly, there is always a point at which the benefits of a new technology outweigh the risks and costs, and that point can be quantified with proper analysis (fig. 1–7).

When it comes to smaller capital expenditures, one way to hedge the technology curve is to rent, rather than buy, equipment. One major supplier of drilling tools now earns about half of its revenue from rentals, as buyers are increasingly seeing the opportunity to lower cost and shift technology risk to the supplier.

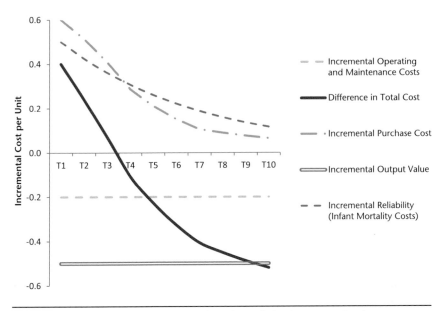

Fig. 1–7. Illustrative incremental cost-benefit analysis of a new technology, over time

Materials Unavailability Strategies

Raw material price risk has increased in recent years. As just one example, metals prices have fluctuated wildly, as shown in figure 1–8.

Aside from purchasing futures contracts on a mercantile exchange, buyers have four options:

- Use substitute materials
- Plan farther in advance
- Buy or control the source
- Refurbish, recycle, or reuse the material or equipment to conserve it

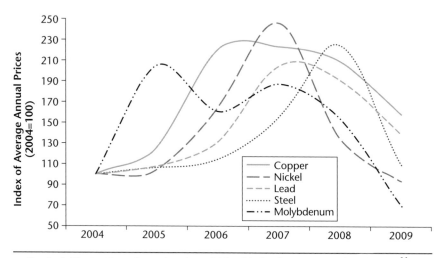

Fig. 1–8. Fluctuation of metals prices, 2004–09 (for first quarter of each year)[11]
(*Source:* U.S. Geological Survey 2005.)

Substitute materials

Just as the automotive industry has transitioned from mostly steel to mostly plastic components to lower cost and weight, specialty chemicals buyers (catalyst manufacturers in particular) have replaced rare earth metals with other materials. Pump, turbine, and compressor manufacturers have replaced specialty alloys with other alloys that are less dependent on materials that are in short supply or in a state of continual depletion. Finding substitute materials takes time so is often the object of an ongoing initiative to ensure that a materials shortage does not interrupt production and a price spike does not cause a budget variance crisis.

Advance planning

Suppliers often accuse operators of waiting until the last minute to order products and services. To their defense, some have attempted to plan farther ahead. Here is a framework for planning ahead more methodically and consistently. By use of master-scheduling logic borrowed from materials requirement planning (MRP) systems, planning for materials in shortage should take place in the demand (planning) time fence (during which manufacturing assets can be redeployed) or the demand time fence (order-cycle time window). If planners wait until the manufacturing time fence (line scheduling), there is a much greater chance of material or component unavailability or high prices. Stretching from a shorter to a longer time horizon allows more flexibility to pursue alternative strategies if available.

Acquiring the source

If a materials shortage appears to be chronic, it may make sense to acquire the source. Vertical integration makes sense when the cost of acquiring the materials through external sources exceeds the cost of procuring them internally.[12] Even when materials can be procured at lower unit prices externally, other factors may justify vertical integration, including the costs associated with searching for sources of supply, negotiating prices, and arranging logistics and transportation.

Conserving equipment or materials

As lead times for new equipment have lengthened, refurbishment has become a fairly popular alternative to buying new equipment. Refurbished equipment is often less expensive and can have a shorter lead time, especially if the supplier builds refurbished equipment to stock.

Recycling component materials has also become more prevalent, as evidenced by the introduction of new recovery processes for rare earth metals on the part of refinery catalyst manufacturers. For example, Grace Davison installed metal traps on its catalyst production lines, which recover about 2% of the total rare earth metal volume used to make the catalysts; Grace Davison is also developing a process to recover rare earths from spent catalysts.[13]

By analogy, enhanced oil recovery (EOR) can be considered as a large-scale refurbishment concept, since depletion has reached the point where recovery processes are widespread.

Project Structure Choices

Solution bundling

The oil, gas, and power industries have typically used a high degree of solution bundling, owing to project complexity, risk, task interrelatedness, and technology intensity. This makes suppliers happy because solution bundling usually allows the supplier to earn higher profit margins; conversely, unbundling often presents a cost-savings opportunity to operators. Operators regularly need to assess the following:

- Whether to buy raw materials separately from components, particularly if the raw material constitutes a very large proportion of the cost of the component (e.g., steel in a large fabrication)

- Whether to buy some components separately from other components (e.g., buying motors separately from the pumps that they drive)

- Whether to buy products separately from the services that are associated with them (e.g., buying compressors separately from their installation and commissioning)

The commoditization of manufacturing has been catalyzed by aggressive strategic sourcing efforts in the automotive industry. In response, suppliers have been struggling to increase their value added, to justify or maintain price levels (fig. 1–9).

Suppliers can protect or enhance their margins most effectively by concurrently combining three strategies: layering on value-added services, building a technology edge, and positioning themselves as the premium provider (fig. 1–10).[14]

Whereas suppliers nearly universally prefer solution bundling, buyers need to be discerning about when to bundle and when to unbundle. Three bundling configurations can optimize value and profit for an operator: complementing, integrating, and solutioning. Complementing adds the least value; integrating adds more value than complementing; and solutioning can offer the most benefit to buyers. Buyers should determine which paradigm best suits their needs and embed the purchase paradigm specifically in requests for proposals and generally in communications with suppliers.

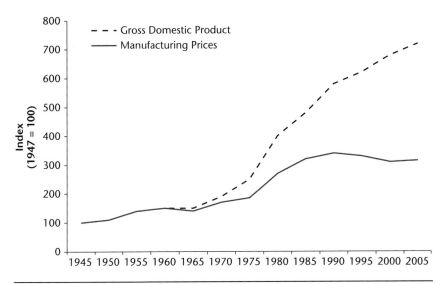

Fig. 1–9. U.S. manufacturing prices since 1945 (*Source:* Author's analysis of data from the Bureau of Economic Analysis and the National Association of Manufacturers.)

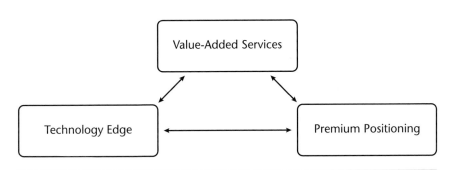

Fig. 1–10. The value creation triangle (*Source:* Jacoby, David, and Bruna Figueiredo. 2008. "The Art of High-Cost Country Sourcing." *Supply Chain Management Review.* June. P. 36).

Level 1: Complementing. The most widespread method of bundling is combining existing product or service offerings so that they complement each other, thereby reducing transactions costs (recall Coase's theory of the firm[15]) for the buyer. Companies offer similar services to serve as a one-stop shop. This combination allows a buyer to reduce the number of suppliers (and, thereby, the cost of managing them). More important, this allows them to increase the quality and effectiveness of the solutions because multiple different products and services should be designed with integration and compatibility in mind. For example, the following complementary bundling solutions should reduce overhead costs for operators:

- Schlumberger's $11 billion acquisition of Smith International allows it to offer products to support the entire drill string (from the bit up to the rotary table on the rig), adding drill bits to Schlumberger's portfolio and giving it full ownership of drilling-fluids market leader M-I SWACO.

- Baker Hughes' acquisition of BJ Services, adding pressure pumping services and equipment to its portfolio, allows Baker Hughes to provide a broader range of services in more geographies than it could before the acquisition, as well as bundling different types of oil and gas production chemicals for which each firm had a strong reputation.

Level 2: Integrating. In integrative bundling, suppliers not only sell the initial equipment or services, but also provide ancillary services throughout the life of the initial product or service. Examples include service contracts, or power by the hour.

One example of this is Flowserve's quick response center, which is part of its LifeCycle Advantage program. The program was created to deliver the lowest possible total cost of ownership by managing spare-parts inventories and maintenance scheduling so as to reduce life-cycle costs, to increase mean time between repair and equipment life, and to maximize uptime (reliability).[16] Another example is SKF's five-year contract with Metso Lindemann to operate its global spare-parts warehouse and distribution operations. SKF now supplies Metso Lindemann's customers with inbound operations, warehousing, customer service, and outbound operations from a central distribution center in Tongeren, Belgium.[17]

Level 3: Solutioning at the equipment level. In solutioning, suppliers not only offer a product and a service but also guarantee a minimum level of performance *for the equipment or service.* Performance-based contracts and value-based pricing seek the maximum product or service performance, rather than the lowest cost. For example, an operator might pay a turbine supplier a bonus if the turbine's output exceeded the rated capacity. In a performance-based logistics (PBL) contract set up by the U.S. Navy, parts availability was as low as 43% under the Navy's management but under GE, also in a PBL arrangement, averaged 90%–100% availability; the PBL contract with GE also helped to eliminate 718 back orders and cut repair turnaround time by 25%. The U.S. Army instituted a similar program for GE's T700 engine, and GE, using lean management, cut repair turnaround times from 265 to 70 days.[18]

Level 4: Solutioning at the system level. Suppliers not only offer a product and a service but also guarantee a minimum level of performance *within the process that the equipment supports*. Thus, the difference between levels 3 and 4 is that while level 3 commitment relates to the equipment or service itself, level 4 relates to the performance of the system in which that equipment is placed. PBL contracts in this configuration provide useful examples for assigning accountability and liability for overall system performance. The U.S. Navy has set up various PBL arrangements for the Rolls-Royce jet engines that it operates. For example, in one case Boeing coordinates with Pratt & Whitney, which makes the F117-PW-100 engine, and other C-17 subcontractors to maximize aircraft availability, the main key performance indicator of the Navy's program.[19]

Success can result in a windfall to the supplier, especially if there is a multiyear contract based on a lump sum and if it obtains prenegotiated financial incentives for higher performance and then hits the targets. In the oil and gas industry, for example, when a bearing manufacturer doubled the price of the equipment, the operator was still happy to pay the extra because the increase in uptime was worth far more than the higher equipment cost. Also, some drill bit manufacturers offer pricing on a cost-per-foot-drilled basis, instead of per bit; consequently, they are charging more on average per job but guaranteeing an increase in rate of penetration,which justifies higher equipment spending owing to the resulting increase in revenue.[20]

Ownership control decisions

At the highest level, the decision about what to do in-house and what to outsource drives most of the cost and risk in a project. Five alternative forms of project structure (see table 1–1) provide a spectrum of choices regarding how much cost and risk to bear and how much to off-load to partners or suppliers.

In the following framework, a project's sponsor is the party that initiates the project. A sponsor can be an oil producer or refiner commissioning a floating production, storage, and off-loading vessel (FPSO), a refinery, a utility building a power plant, or a government entity (e.g., a power commission) that offers concessions for facilities. The contractor is the firm that builds the project, which may or may not operate and maintain the facility once it is completed. The following configurations of ownership may be used to achieve the desired level of engineering, construction, operation, and maintenance risk (roughly in order of most to least in-house responsibility):

- **Design-build-operate (DBO).** The contractor assumes full responsibility for all phases of the project. Maintenance may be outsourced (if it is performed by the contractor, then it is a design-build-operate-maintain [DBOM] contract).

- **Build-own-operate (BOO).** The contractor manages the construction and operation aspects, but takes no responsibility for the design phase. Again, maintenance may be handled by the operator or is outsourced.

- **Build-own-operate-transfer (BOOT).** The contractor takes full responsibility for the project and then sells the resulting installation. This sets the contractor up to be a repetitive builder of whatever the product is (e.g., FPSOs or offshore wind farms).

Sponsors can also shift risk to third parties by hiring an EPC or EPCI firm to manage the project. Then, by structuring the contract with the EPC or EPCI firm, the sponsor can achieve the desired level of risk during the construction phase. Here are some options (in order of most to least in-house responsibility):

- EPC/time and materials (T&M)
- EPC/cost-plus

- EPC/LSTK
- EPCI/T&M
- EPCI/cost-plus
- EPCI/LSTK

The complexity of these arrangements can be simplified into five basic types of agreement, as summarized in table 1–1.

Table 1–1. Assignment of risks assumed by EPC contractors under five types of agreement

	DBOM	BOO	BOOT	EPC(I)/ LSTK	EPC(I)/ Cost Plus or EPC(I)/T&M
Ownership of site	Contractor	Contractor	Contractor/ Sponsor	Sponsor	Sponsor
Acquisition cost risk	Contractor	Contractor	Contractor	Contractor	Sponsor
Design risk	Contractor	Sponsor	Sponsor	Sponsor	Sponsor
Operations and maintenance cost risk	Contractor	Contractor	Contractor/ Sponsor	Contractor/ Sponsor	Sponsor
Prime contractor performance risk	Contractor	Contractor	Contractor	Contractor	Sponsor
Change orders	Contractor	Contractor	Contractor	Contractor	Sponsor
Project management resources	Contractor	Contractor	Contractor	Contractor	Sponsor
Contractor risk index	▃▅▆	▃▅▆	▂▄▆	▂▃▅	▁▂▄
Sponsor risk index	▁▂▄	▁▂▄	▁▃▅	▃▄▆	▃▅▆

Make versus rent versus buy criteria

The *make versus buy* decision drives cost in the same way as the ownership control decision discussed above.

In addition, deciding whether to rent versus buy can similarly change the risk/return profile of a supply chain activity by outsourcing part or all of it. Outsourcing affects not only risk profile but also cost, effectiveness, and sustainability. Potential reasons for outsourcing include the following:

- *Lower cost and capital requirements.* For instance, one production chemical supplier does not make any chemicals. It buys the base chemicals, mixes them, and resells them to the oil producer or service company. This allows it to avoid tying up capital in facilities and gives it the flexibility to choose the best supplier for a given type of chemical without developing the chemistry itself.

- *A more global network, including enhanced access to foreign markets.* For example, one supplier of drilling and completion fluids outsources production of most component chemicals to local suppliers, to reduce shipping time and cost. Instead of investing in capacity in each region, it has developed a network of local suppliers. As most of the lead time for chemicals is related to shipping, outsourcing allows this supplier to deliver faster than competitors that produce in another region. In one instance, it arranged next-day delivery of a drilling-fluid additive, sourced from a local partner, when the supplier with the next shortest lead time quoted a week.

- *More sophisticated and updated information technology platforms that provide higher accuracy and reliability compared to in-house systems.* Many companies outsource logistics primarily because the electronic manifests and other supply chain visibility systems cost too much to create and update in-house. Third-party logistics providers continually invest in new information technology because they are serving a multitude of customers and can pass on a fraction of the cost to each one.

- *Value-added services that would be challenging or expensive for the operator to do in-house.* Often, the cost of in-house staff makes value-added logistics tasks more expensive than using outsourced providers' equivalent services—for example, kitting, light assembly, mixing, blending, and similar customization operations that are performed on a finished or semifinished product. Outsourcing these activities can often take advantage of the lower cost structure of an operation that is set up with non-union, part-time, or offshored labor to perform them at a lower cost than would be possible with in-house personnel.

In consideration of whether to outsource an activity, it is important to evaluate the loss of skills and talent that will result. The loss of know-how could be permanent if the activity cannot easily be reconstructed in-house. One turbine user employed the following criteria to determine whether maintenance should be outsourced:

- Does the process represent a core competence that should remain in-house even if there is a cost disadvantage?

- Is it less expensive to outsource the process, including all in-house and outside life-cycle costs?

- If the process is more expensive to perform in-house, could the in-house process be engineered differently to reduce the cost enough to meet the outside cost?

- Would an investment be required in order to outsource the activity, and if so, would it pay off? Conversely, if outsourcing would allow equipment and tools to be liquidated, would any value be recaptured from selling the assets?

- Could the activity be brought back in-house in the future if the outsourcing experience is deemed to be unsuccessful?

- Does management have enough time to manage a transition from in-house to outside, or vice versa?

- Would renting equipment or asking the vendor to offer life-cycle solutions provide a viable alternative to outsourcing? If so, would this affect the pace of innovation or efficiency gains?

Notes

1 Shively, Bob, John Ferrare, and Belinda Petty. 2010. *Understanding Today's Global LNG Business*. Laporte, CO: Enerdynamics. P. 97.

2 Jacoby, David, and Erik Halbert. 2007. Energy prices re-shaping the supply chain: Charting a new course?" Boston Strategies International. http://bostonstrategies.com/images/BSI_-_SS4_Energy_Prices_Reshaping_the_SC.pdf (accessed May 19, 2012).

3 $P_n = P_0 \times [W_1 \times (F_1/F_{1-0}) + W_2 + (F_2/F_{2-0})]$, where P_0 represents the original negotiated price (at time 0), W represents weighting factors/percentages of alternative fuels, F_1 and F_2 represent alternative fuels. An inflation component may be added. See Tusiani, Michael D., and Gordon Shearer. 2007. *LNG: A Nontechnical Guide*. Tulsa, OK: PennWell. P. 329.

3 CAPS Research (Institute for Supply Management). 2011. *Petroleum Industry Supply Management Performance Benchmarking Report*. Tempe, AZ. P. 8.

4 Ibid.

5 "High-Impact Sourcing: Precision-Guided Sourcing Strategies for Maximum Results." 2005. Boston Strategies International (formerly Boston Logistics Group) study.

6 Boston Strategies International study. 2009; Jacoby, David. 2010. The oil price "bullwhip": Problem, cost, response. *Oil & Gas Journal*. March. Pp. 20–25.

7 Jacoby, 2010, pp. 20–25.

8 Kizis, Joseph A., Jr. 2009. Bravo options barite rights to Baker Hughes. Homestake Resource Web site. http://www.bravogoldcorp.com/en/news/147/bravo-options-barite-rights-to-baker-hughes.php (accessed May 19, 2012).

9 Jacoby, David. 2010. Using flexible capacity techniques to thrive in a volatile economy. *Logistics Digest*. February 9. P. 31. http://www.logisticsdigest.com/inter-education/inter-opinion/item/4669-using-flexible-capacity-techniques-to-thrive-in-a-volatile-economy.html (accessed May 19, 2012).

10 Kuwait Oil. [n.d.] General Conditions of Contract for Lump Sum Turnkey Projects. http://www.google.com/url?sa=t&rct=j&q=register%20of%20commerce%20no.%2021835&source=web&cd=1&ved=0CCMQFjAA&url=http%3A%2F%2Fmcsetender.kockw.com%2FCommercial%2520Documents%2Fstandards%2FGCC%2520LUMP%2520SUM%2520TURNKEY%2FGCC%2520LSTK%2520%2520March%25202007.pdf&ei=gCY4T5vlEczq0QGyma28Ag&usg=AFQjCNH0CsdnTWzz3BngjwsnWa5OYGyOqQ&cad=rjt (accessed May 19, 2012).

11 U.S. Geological Survey. 2005. Data Series 140. http://minerals.usgs.gov/ds/2005/140/; Metals Consulting International Limited [n.d]. http://www.steelonthenet.com/kb/steel-billet-prices-1998-2009.html; Metals charts and nickel forecasts. [n.d]. Stainless Steel Information. http://www.estainlesssteel.com/zmetalsandstockcharts.shtml.

12 Coase, Ronald H. 1937. The Nature of the firm. *Economica*. 4 (16): P. 386. http://www.sonoma.edu/users/e/eyler/426/coase1.pdf.

13 Grace, W. R. 2011. 2011 Q1 earnings call.

14 Jacoby, David, and Bruna Figueiredo. 2008. The art of high-cost sourcing. *Supply Chain Management Review*. May/June. P. 32.

15 Coase, 1937.

16 *Fair Disclosure Wire*. 2011. Q1 2011 Flowserve Corp earnings conference call—Final FD. April 28.

17 SKF. 2010. SKF signs contract to operate warehouse and distribution for Metso Lindemann. Press release, May 5. http://www.skf.com/portal/skf_gb/home/news?contentId=886178.

18 Innovative Strokes for Readiness. 2008. *Overhaul & Maintenance*. July 1. P. 3.

18 Ibid.

20 Rabia, H., and M. Farrelly.1986. A new approach to drill bit selection. Paper presented to the Society of Petroleum Engineers. European Petroleum Conference, London, October 20–22. http://www.onepetro.org/mslib/servlet/onepetropreview?id=00015894 (accessed May 19, 2012).

2

ENGINEERING AND PROCUREMENT OF EQUIPMENT AND SERVICES AT MINIMUM TOTAL COST AND RISK

Bid Slate Development and Supplier Selection

Early choices in bid slate development influence the cost of the project, not only by narrowing the choice of suppliers, but also by signaling to suppliers the competitive dynamics of the ensuing process of bidding, negotiating, and contracting. Key decision points involve the determination of the number of suppliers and the short-listing of preferred ones.

Category management

The first step in bid slate development is implementing a robust process for category management and organization. Category management helps to mitigate supply risk by establishment of formal relationships with suppliers in categories of goods and services that are of strategic importance to operations.

Category management is a process that defines the desired end state of procurement activities for each family of externally purchased goods or services. It integrates the processes of supply risk management, supply market analysis, and market intelligence to choose how to maximize value from suppliers under prevailing market conditions. Its scope includes the selection of tendering and request processes, bid slate development, bid comparison methods (e.g., total cost of acquisition evaluation methods), risk profile of contracts (e.g., LSTK vs. cost-plus), the manufacturing strategy (engineered-to-order vs. made-to-order [ETO and MTO, respectively]), and which value-added services should be included in the contract. Through these definitions, management can prepare a strategic negotiating position that helps to achieve the mutual goals.

Steps in establishing a category management program include, for each category:

- Analyzing external expenditure
- Understanding the supply market dynamics
- Formulating and executing a strategy to get the most value from suppliers
- Providing cross-functional career paths that develop people for category management roles and offer them an opportunity to apply their knowledge after they progress to another place in the organization

Appendix B lists commonly sourced commodities and categories thereof, to help in establishing category management families. These are groups of externally purchased goods and services that may be addressed within a common set of goals, supply chain strategies, and sourcing tactics.

Determining the optimal number of suppliers

In the procurement of engineered products, having fewer suppliers is generally more effective than having more suppliers. Single or dual sourcing can raise concerns about the competitiveness of the marketplace and of the sourcing event. However, an analysis of industry concentration and prices[1] has shown that there is no correlation between supplier concentration and price inflation (fig. 2–1); moreover, empirical research[2] has shown that, even in anticompetitive situations, reductions in cost through synergies more than offset anticompetitive increases in prices. There is also empirical evidence that price is typically determined by the top two bidders, and does *not* decline with more than four bidders.[3] Therefore, single sourcing is often a viable option for services that are time critical, if supply market capacity and in-house capabilities are limited.

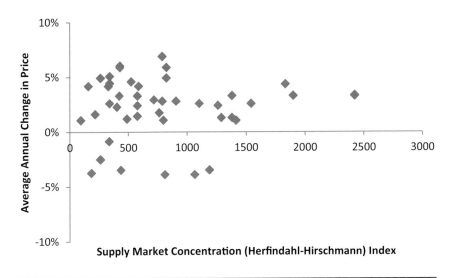

Fig. 2–1. Correlation between industry concentration and price inflation for
selected oil, gas, and power supply markets, 2007–10 (*Sources*: Boston
Strategies International analysis, data adapted from U.S. Bureau of
the Census.)[4]

Unless there is price signaling or a breakdown in the bidding
process, suppliers will bid competitively even if there are only a few
bidders. So long as there are at least four suppliers in the market,
the market is typically a competitive one. This interpretation
is consistent with legal precedent for determining the level of
concentration that constitutes monopolistic market conditions
under which a dominant supplier may exert an influence over prices.
Although the law is complex, varied, and subject to interpretation,
assessments[5] of unacceptable concentration have ranged between
about 25% and 35% market share for an individual supplier, which
would simplistically equate to three or four suppliers in a market.

In fact, suppliers may even underbid (below their cost) to win
projects, which should be a concern to buyers insofar as it may
impair the suppliers' viability. Excessive supplier competition has
driven some unsophisticated suppliers, who don't know their
actual cost to serve, out of business, as they over-react to extremely
competitive situations such as time-based reverse auctions. That
being said, the savings from competitive events are extraordinary.
On average, companies have saved 6%–8% on their largest

categories of expenditure.[6] Best-in-class companies save 10%.[7] For individual companies and specific categories, there are many cases of savings in the 30%–75% range.

Even if having few suppliers results in higher margins paid to those suppliers, sophisticated buyers will get more than the differential out of the relationship in the form of partnering benefits such as collaborative planning, inventory consolidation, and joint process improvements. Anticompetitive behavior can raise prices by 3%–4% in concentrated markets, but efficiency gains that come from lower transactions costs, economies of scale, and better customer service typically outweigh that, resulting in a net 1% price decrease.[8] Electronic procurement (e-procurement) alone, which is implemented more effectively in close collaboration with partner suppliers, can save both buyer and supplier by elimination of labor-intensive paperwork and reconciliation of discordant data elements. Qatar Petroleum, the Bahrain Petroleum Company (BAPCO), and the Kuwait National Petroleum Company (KNPC) use an external provider for spend analysis, strategic sourcing, and electronic requests.[9]

Single and dual sourcing are both viable options, depending on the situation (fig. 2–2). For a large expenditure on a relatively simple product or service, given no time or capacity constraints, normal market concentration, and many competitive options, operators would select a multisource strategy (three or more suppliers). However, if the market is capacity constrained and there is little supplier risk, dual sourcing is not unreasonable. Furthermore, if a strategic target supplier is not weak in critical areas (financial vulnerability, research and development capability, etc.), if switching costs are not too high, and if the supplier's price quote is consistent with benchmarks, then single sourcing could be a reasonable option as well.

Highly concentrated markets—those above 1,800 on the Herfindahl-Hirschman index—place a special responsibility on buyers to develop the market and to ensure price competitiveness. Overly concentrated supply markets raise the possibility of certain types of illegality, such as vertical or horizontal price fixing, price signaling, predatory pricing, price discrimination, promotional discrimination, and geographic discrimination (carving up markets).[10]

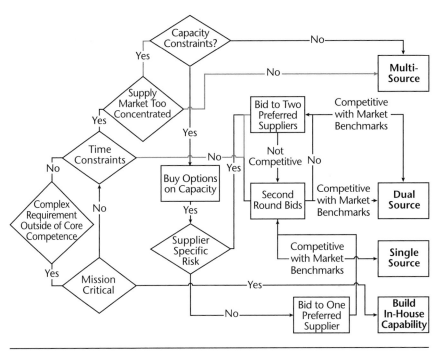

Fig. 2–2. How to decide whether to single or dual source

Tools for buyers to avoid overpaying in highly concentrated markets with few suppliers include the following:

- **Dual sourcing.** Dual sourcing can maintain a competitive option and avoid high switching costs. In particular, developing smaller or less-qualified suppliers can provide options for future development on technically or regionally specialized projects or on a divisional basis.

- *Should-cost modeling.* This is a process for determining what a product should cost on the basis of its component material costs, manufacturing costs, production overheads, and reasonable profit margins. (A detailed example of should-cost analysis is included in chap. 5.)

- *New-supplier development.* Buyers should continuously develop new and alternative suppliers to avoid high potential switching costs. Supplier development is a long-term exercise and often yields unexpected innovation and cost reduction opportunities from new potential suppliers. Supplier development activities should first include consideration of existing suppliers of related products or services. Global sources in strategic or target markets can also serve as good benchmarks and potential complements to or substitutes for entrenched suppliers.

- *Legal recourse.* Whenever evidence or suspicion of price fixing surfaces, buyers can pursue legal recourse for suspected price fixing or related crimes. Large expenditure categories have historically seen their share of wrongful activity. For example, in 2007, Siemens, Schneider Electric, Areva, Alstom, Mitsubishi Electric, Hitachi, ABB, and four others were convicted, by the European Commission (EC), on grounds of fixing prices for electrical switchgear for 16 years. The EC imposed the second-largest fine in its history to date on the firms—a total of $977 million—although ABB ultimately escaped the fines because it had reported the offense to the EC.[11] Also, in 2010, the Japanese Fair Trade Commission fined 10 companies, including Sumitomo, Furukawa, and Fujikura a total of $180 million for setting prices of fiber-optic cables and components that they had sold to Japanese telecommunications firm Nippon Telegraph and Telephone.[12] In a preemptive move, the U.S. Department of Justice imposed conditions on the merger of Baker Hughes and BJ Services' operations in North America. Before completely integrating, Baker Hughes was compelled to sell two stimulation vessels, as the merger would have combined two of only four companies providing offshore well stimulation services in the Gulf of Mexico.[13]

When an empirical basis is preferred to decide how many suppliers should be approved or engaged, a decision analysis tool can be used. For extremely large expenditure decisions, this author has helped clients determine the optimal number of suppliers on the basis of an analysis of the probability of occurrence and the monetary severity (and the probability distribution of each of those) of five undesirable outcomes at varying levels of number of suppliers. The five outcomes are (1) cost of nonstandardization and the risk of (2) price gouging, (3) output unreliability, (4) missed delivery dates, and (5) project management error.

Qualifying suppliers

Supplier qualification is an especially critical job in oil, gas, and power generation. Projects are time driven, and the nonperformance of even one supplier may delay a whole major project, for example a refinery coming onstream, at very high opportunity cost. Also, safety and environmental risks are high, so supplier quality lapses could be life threatening. One mechanical failure could cause a rig blowout (e.g., as occurred at Macondo field), and one safety violation could cause an explosion (e.g., faulty seals in electrical conduit in a classified area). Therefore, qualification is typically a multistage and rigorous process involving the following steps, which define key success factors for suppliers:

- Prescreening
- Prequalification as an approved vendor
- Qualification for specific projects and applications
- Technical evaluation
- Approved-vendor nonbinding contract
- Selection for a project or application
- New supplier/project registration and on-boarding

Prescreening. Prescreening typically involves a review of the company's technical expertise, experience, and financial strength. At this stage, the financial assessment often focuses on total revenue, profitability, and solvency, sometimes with the use of financial ratios like the quick ratio. A supply market intelligence program can also help prescreen vendors in a systematic way—for example, targeted at certain geographies and product or solution offerings.

Since some large investment decisions are based on the availability of suppliers to maintain equipment for 20 years or more, procurement staff need to assess their performance on the key success factors that determine long-term viability. For example, key success factors that define the success of major oil field service and construction firms include: (1) demonstrated technological innovation; (2) presence and history in the major oil and gas markets; and (3) international network (especially in the fast-growing areas such as the Middle East and Asia).

Consider how the following leading players demonstrated the above attributes early in their development, laying the groundwork for their rise to become today's dominant suppliers:

- Schlumberger was the world's first well-logging company. In 1929, the company carried out subsurface surveying in Argentina, Ecuador, India, Japan, the Soviet Union, Venezuela, and the United States; furthermore, Schlumberger performed the United States' first well log that same year. In 1940, the company moved its headquarters to Houston to be close to its oil and gas clients.

- Halliburton performed its first offshore cementing job using a barge-mounted cementing unit at a rig in the Creole Field in the Gulf of Mexico, in what was to become the most extensive offshore service job in the world at that time. It opened offices in Canada in 1926 and Venezuela in 1940, and by 1946, the company was operating in the Middle East and had begun working for the Arabian-American Oil Company (the forerunner of Saudi Aramco).

- Weatherford pioneered an innovative technique and equipment for cementing cased-hole oil wells starting in 1941. It expanded in Venezuela when the Gulf Oil Company, which invested in Venezuela in 1943, began using its products.

- National Oilwell Varco's (NOV) predecessors introduced the first rotary drilling table in 1913. NOV now has a distribution network of over 700 locations around the world.

- Saipem (Italy) started offshore operations in the 1960s. It developed facilities in West Africa and the Middle East by leveraging local content. A number of acquisitions culminated in the acquisition of Bouygues Offshore in 2002.

- Fluor (U.S.) won a contract to build a natural gas plant for Richfield Oil in 1922 and located in California, where the company got its first major refinery contract for Shell Oil in 1932. Fluor went on to secure its first major overseas contract for Saudi Aramco in Saudi Arabia in 1947.

Order lead times can also serve as a barometer of how much available capacity suppliers have, which can be a prescreening criterion if available capacity is a consideration for preselection (fig. 2–3).

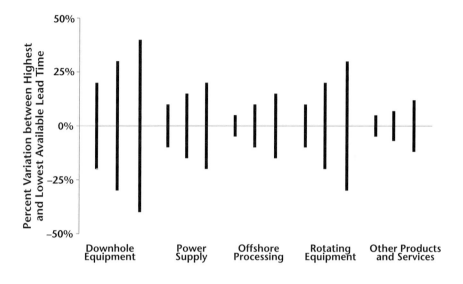

Fig. 2–3. Range of lead times from different suppliers for the same product or service at the same time, 2008–09 (illustrative)

Prequalification. Granting approved-vendor status is a multistage process (fig. 2–4). A sales meeting is often followed by a request for a more detailed package, including information that may be requested at various points in time, such as

- Basic company profile
- Corporate size/financials
- Organization
- Proposed facilities
- Order management procedures
- Order management systems
- Shipping and receiving procedures

- Safety procedures and safety records
- Administrative controls
- Performance along key performance indicators
- Technology
- Reporting/visibility tools
- Customer reference checks
- Pricing

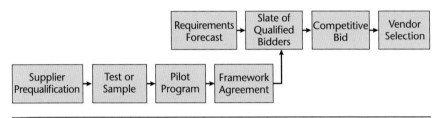

Fig. 2–4. Simplified representation of a supplier qualification process

Technical evaluation. Product samples, laboratory testing, an on-site quality visit, and a pilot project may be involved in evaluating a supplier's technical capabilities and competence. Sometimes the visit is conducted by an independent certifying body or a consulting firm with an approved auditing methodology. Additional factors considered at this stage may include

- Technical capability and fit
- Potential operating-cost savings
- Value addition as a supply chain partner

Approved-vendor. If the vendor is approved on technical grounds, a commercial framework agreement is often signed, which locks in terms and conditions so that future bids can proceed with less negotiation on these aspects. Basic commercial terms and conditions covered in this contract would include

- Alteration of terms
- Assignment

- Audit rights
- Changes
- Confidentiality
- Delivery: schedule and delays
- Force majeure
- Governing law
- Independent contractor usage and subcontracting
- Inspection
- Insurance and indemnification
- Intellectual property: design, data, creative work, patents, and inventions
- Liens
- Payment—including discounts and refunds
- Price structure
- Safety
- Scope
- Service and/or spare parts (if applicable)
- Severability
- Substitution
- Taxes
- Term
- Termination (for cause and for convenience)
- Warranty

Preselection for a project or application. For specific requirements, the procurement department and the user (operations, engineering, etc.) usually ask a subset of the approved vendors to bid on the required services.

Contracting. Once a supplier is technically approved, it becomes eligible to participate in tenders, and if it wins a bid, it may be necessary to further qualify the exact products to make sure it delivers the desired results (fig. 2–5). The following steps may be necessary for this evaluation:

- Determine which items need samples and get samples
- Set up testing program and timetable
- Test product and approve new sources
- Get Material Safety Data Sheets (MSDSs)
- Register and update vendor information in appropriate ordering system
- Ensure item numbers are updated and mutually exclusive
- Create blanket orders for new suppliers
- Establish controls against maverick spending

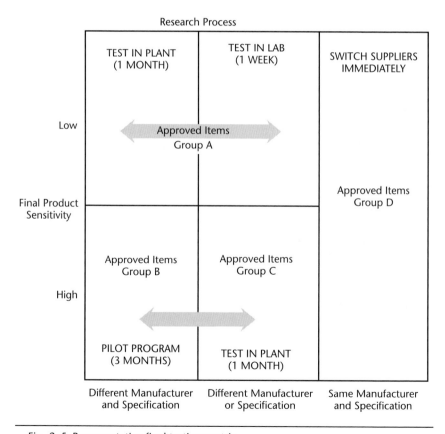

Fig. 2–5. Representative final testing matrix

- Communicate any changes of suppliers to users
- Communicate sourcing decisions to plants when all else is finalized
- Establish delivery frequencies and time windows, by plant
- Check/update stocking locations, space, and inventory levels per delivery frequencies, by plant

Low-cost country sourcing

Are low-cost country sources qualified? Oil and gas companies are generally not keen on low-cost country sources because of reliability concerns. However, their suppliers and their suppliers' suppliers use them routinely. The share of oil field equipment coming from Asia has increased, from 34% in 2007 to 46% in 2012, and can be expected to increase to nearly 50% by 2013 (fig. 2–6).

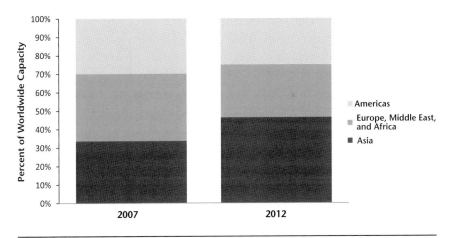

Fig. 2–6. Change in oil, gas, and power industry supply market capacity by region, 2012 vs. 2007

Cheap labor is the first reason many suppliers moved sourcing efforts to low-cost countries, chiefly China. For a long time, China's labor rates were about 4% of those found in Belgium, Germany, and the United States, which are the most expensive nations, and were lower than those of Mexico and India. A 2008 study[14] found that the average cost of Chinese labor was as low as 10% of the cost of U.S. labor. Even after accounting for the approximately 20% of product cost that was dedicated to transport, Western manufacturers recently enjoyed as much as a 70% cost advantage by sourcing components from China.[15] Second, lenient regulatory controls have made low-cost countries attractive sourcing platforms by lowering their costs relative to those of their Western counterparts. Third, China put unprecedented investment into its infrastructure, thereby addressing some of the conditions that had previously made it unattractive from a logistical point of view. Finally, it joined the World Trade Organization, established accepted legal frameworks that moved toward protecting intellectual property, and trained workers.

Recently, the price difference for sourcing from China in particular has eroded somewhat owing to rising labor costs in China, environmental legislation in China, and the economic crisis, which has caused some manufacturers to relocate production back to their home countries. Labor costs nearly doubled in urban areas like Shanghai in the space of a few years, eliminating about half of the cost advantage of sourcing from China (going from 10% to 20% of U.S. costs, leaving only a 60% cost advantage for U.S.-based companies to source from China after accounting for transportation). While Chinese labor rates increased 7%–10% in 2005, U.S. wages rose by about 4%, before adjusting for inflation. Salaries rose 7%–9% in cities such as Beijing, Tianjin, Shanghai, and Guangzhou, according to the U.S.-China Business Council.[16] Wages in smaller cities, such as Chengdu, Hangzhou, and Wuhan, rose by the same percentage or more. The more recent support on the part of China's government of a doubling of the minimum wage ensures that this trend will continue.

China's gradual revaluation of its currency, the renminbi, against the U.S. dollar—which made Chinese goods 5% more expensive in 2009 than they were in 2008—is also eroding China's cost advantage.[17] In addition, Chinese authorities raised export taxes on certain raw materials by about 12% by eliminating an export credit, which further reduced the cost advantage to about 30%–35% for these items.

Many oil companies perceive low-cost country sourcing to constitute an unacceptable quality risk, especially for complex equipment, such as turbines. However, suppliers have taken the low-cost route without waiting for their customers' approval. While the low-cost wave started in about 1985, with pulleys and fan belts from China, today even high-technology companies are heavily sourcing and manufacturing there. For example, Rolls-Royce signed a licensing agreement with STX Engine, in South Korea, for Rolls-Royce's RB211-H63 gas turbine, which is designed for offshore platforms, FPSOs, and drill ships.[18]

India, Brazil, and Russia also present increasingly attractive sourcing locations for importers in the Middle East, North America, and Western Europe. Champion, Nalco, and Huntsman increased their market shares in India by acquiring rival companies or establishing research facilities. Huntsman bought the chemicals division of Laffans Petrochemicals, giving it a production base in India, Nalco opened an $8.5 million research and development and service center in India, and Champion bought 50% of Basic Oil Treating India from Dai-Ichi Karkaria, a maker of specialty chemicals for oil and gas production.[19] In addition, the growth of new emerging economies, such as Vietnam, Thailand, Hungary, and Turkey, will offer ongoing opportunities for low-cost sourcing.

Establishing and Maintaining Partner Relationships

Once a supplier is selected, operators need to decide how to allocate resources to get the most out of the supplier. This starts with clarifying the degree of intimacy that will lower the cost of communication (and associated miscommunications and omissions) and increase the effectiveness of the combined resources by leveraging synergies.

Structuring alliances

A comprehensive partner supplier program guides supplier relationships from the commodity stage through five levels of partnering toward strategic partnership:

- *Transactional.* Suppliers take orders as needed from the operator.

- *Preferred.* Suppliers have met some prespecified conditions, such as a vendor or quality certification, that distinguish them from the transactional suppliers.

- *Value-added.* Suppliers have a framework agreement to provide a focused service.

- *Alliance.* Suppliers participate in joint development efforts with the buyer, such as oil contractor Halliburton has with many of the major oil companies in the development of oil fields.

- *Strategic partnership.* Suppliers have established a common vision based on mutual needs and strategies. Companies that form true partnerships collaborate intimately with suppliers to achieve not only lower total cost but also faster speed to market, greater innovation, and better quality.

There are nine dimensions to partnerships. On each of the nine dimensions, the depth and nature of the partnership evolves as the relationship progresses through the five rungs of the partnership ladder, as shown in figure 2–7.

Prequalification for a commodity supplier can consist of a simple, Web-based application process. As the relationship progresses to a preferred supplier, site visits are required to understand the operating environment. When considering the same supplier for a value-added role, that supplier's past track record with the operator should be assessed. Alliance partners need to be approved at the division or business unit level, and strategic partners must be implicitly or explicitly blessed by executive management or board of directors. Performance measurement escalates from being a one-time, project-based task to an ongoing and periodic assessment as the relationship becomes more symbiotic (fig. 2–8).

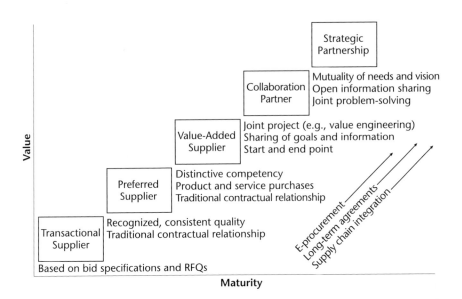

Fig. 2–7. The Partnership Ladder (*Source:* Adapted from Michael Maccoby and Jacoby, 2009.)

Level Number	1	2	3	4	5
Partnership type	Transactional	Preferred	Value-added	Alliance	Strategic partner
Selectivity	None	Top 50%	Top 25%	Top 5%	Top 1%
Qualification procedure	Information exchange	Site visit	Joint track record	Divisional approval	Board vote
Formal agreement	Purchase order	Contract	Memorandum of understanding	Alliance agreement	Joint charter
Points of contact	Account representative	Account executive	Key account executive	Multiple levels	All levels and geographies
Joint planning time horizon	None	Annual	Short-term capital budget cycle	Medium-term capital budget cycle	Long-term capital budget cycle
Frequency of communication	Project-based	Annual	Monthly	Weekly	Daily
Joint investment	None	None	Project office	Semi-permanent colocation	Joint/shared assets
Joint R&D/ I.P. sharing	None	None	Informal	Access rights	Co-development
Performance measurement	Project-based	Periodic	Cumulative	Integrated	Perpetual

Fig. 2–8. Partnership maturity model (*Source:* Author, after Jacoby, David. "On the Cutting Edge of Strategic Sourcing," April 2004; Jacoby. "High-Impact Sourcing," 2005; Jacoby. "The New Economics of Partnering." *Purchasing.* Jan. 13, 2005. P. 60. Also see Lambert, Douglas. "So You Think You Want a Partner," *Marketing Management*, Summer 1996, Vol. 5, No. 2, p. 39.)

Supplier certification or other recognition is often denoted by internally defined thresholds, and the level of certification can increase as suppliers become more strategic, creating *bands* of supplier capabilities. For example, one operator has defined bronze, silver, and gold bands of performance.

Communication frequency increases and management meetings become more frequent, often a natural by-product of the size of the projects. This correlates with increased feedback and joint planning and often requires the nomination of dedicated full-time liaisons from each organization.

Alliance and strategic partners colocate facilities to increase communication and economize on facilities and overhead costs. Joint investments are considered and implemented, including joint research, development, and licensing of intellectual property. In a strategic alliance, the partners typically share equity in a joint-venture organization, which is sometimes driven by local content obligations.

The opportunity presented by partnerships can put a lot of pressure on suppliers to climb the ladder (see fig. 2–7). While managing the supply base as it progresses through these stages, the buying organization must respect ethical norms, avoiding practices such as issuing suppliers requests for quotes to benefit from free engineering while not intending to hire them (the ISM calls this *sharp practices*); sharing suppliers' prices with other suppliers; and accepting gifts or favors from suppliers.

Setting up joint ventures and taking equity stakes

Joint ventures should only be considered when the governors of both businesses determine that the strategies of both firms (perhaps within a certain market space) are sufficiently aligned that the two would be more powerful by merging than competing. Large joint ventures can increase the business risk (the *beta*, as it's referred to in equity analysis) completely independently of offering any supply chain benefits or synergies. Therefore, when considering a joint venture, first ask whether the desired state could instead be achieved through an arms'-length transaction involving licensing or shareholding in each other's publicly traded stock.

Often, conventional contractual or trade-based relationships can be adapted to achieve the desired result without the higher beta value that would accompany a business integration. In contrast, once a joint venture is formed, it can be costly and potentially litigious to dissolve if the relationship goes awry. Following are several examples of international licensing agreements between major suppliers and their international affiliates:

- GE has licensed its centrifugal compressor design to Bharat Heavy Electricals (BHEL) in India, and in 2010, they extended their licensing agreement by 10 years. BHEL manufactures and assembles the compressors in India for sale in India and other South Asian countries where local content and cost are deciding factors for contract awards.[20]

- Also in India, Lufkin licensed its gearbox technology to domestic steam turbine manufacturer Triveni in 2011. The agreement covers gearboxes for compressors, steam turbines, and pumps and gives Lufkin royalty income and an installed base in markets that give preference to local suppliers of technology to domestic steam turbine manufacturers.[21]

- Pump manufacturer Ruhrpumpen gained a strong foothold in the Chinese petrochemical market in 2011 by licensing LianDi Clean Technology as a distributor. LianDi provides engineering services for petrochemical companies in China, and gives Ruhrpumpen an established customer base and familiarity with the local market.[22]

- GE Oil & Gas signed a long-term technology transfer and license agreement in 2010 with JSC ZKMK (Zapadno-Kazakhstanskaya Mashinostroitelnaya), a Kazakh precision-engineering plant operator. The deal makes GE more competitive in Kazakhstan's compressor market, giving it a stronger position to bid on projects there.[23]

- In the market for microturbines, Capstone has qualified over 90 distributors worldwide to sell and repair its equipment, which has helped it expand into new regions and market segments.

Nonetheless, joint ventures are a common form of supply chain integration in countries where legal requirements and deeply embedded procurement practices favor domestically incorporated companies, such as China. Weir Pumps and Mitsubishi Heavy Industries (MHI) formed joint ventures in China in 2010 and 2011 to gain access to Chinese buyers. Weir formed a joint venture with Shengli Oilfield Highland Petroleum Equipment to sell high-pressure oil field pumps in China. The firms will leverage Highland's relationships with CNPC and Sinopec to sell Weir's fracturing pump technology to Chinese shale gas operators.[24] MHI created a compressor manufacturing joint venture with Hangzhou Steam Turbine & Power Group Company, a Chinese firm that makes compressors for petrochemical plants, in early 2011. The joint venture gives MHI a stronger sales presence in mainland China, and access to a lower-cost manufacturing base than its Japanese facilities. However, MHI supplies key components such as bearings from Japan to ensure quality and maintain control of the intellectual property.[25]

In the offshore wind power business, Dong (the Swedish utility) and Bladt Industries (the Danish offshore foundation fabricator) formed a joint venture to ensure adequate capacity for its growing demand. Dong agreed to buy 600 foundations from Bladt's Aalborg plant. The two organizations cooperated on design and manufacturing.[26] Also, SSE, the U.K. utility, took a 15% stake in BiFab, a supplier of foundations. In exchange for the equity infusion, BiFab agreed to provide SSE the option to order up to 50 foundations per year from 2014 for 10–12 years. Both the Dong joint venture and the SSE equity stake clearly offer these operators increased supply chain flexibility in a market that may get overheated.[27]

If management decides to form a joint venture, potential joint venture partners should discuss operating models while forming their mission and governance structures. Within joint ventures, the operating paradigm can favor dominance by one of the partners, or alternatively it can be operated as a truly independent entity exercising control over its own destiny in the model of WellDynamics, a smart-completion supplier owned jointly by Halliburton and Shell. The degree to which each partner manages operations in a joint venture should reflect its experience and success in operating similar businesses, its strategic assets, and its access to customers of joint interest.

The Tendering Process—Choices and Best Practices

Tendering involves five subprocesses:

- Work scoping, especially for services procurement
- Requests for information, proposals, or quotations (collectively known as *RFx* processes, and individually as RFI, RFP, and RFQ), including e-procurement
- Bid analysis, including total life-cycle cost analysis
- Negotiation of strategic and nonstrategic categories
- Contracting of high- and low-value purchases

In contrast to the apparent simplicity of the tendering process as practiced in other industries, the complex nature of projects in oil, gas, and power requires careful consideration of tendering options that can dramatically affect the resulting solution and price. For example, the following should be explicitly discussed and decided:

- How many suppliers to interact with at various points in the tendering process (when and how many to short-list). Inviting too few suppliers to the bidding may result in a narrower solution set and a higher price. On the other hand inviting too many suppliers may result in higher tendering costs, longer time in the selection process, and higher long-run supplier management costs.

- How many rounds of proposals to request (zero, one, or two) and how much design and development work to request or expect from suppliers prior to each milestone. Conducting too few rounds of bidding may result in an incomplete specification and a higher price, while conducting too many may result in higher costs and delay the project.

- When to discuss and negotiate price (early, later, or final). Discussing price too early may compromise innovation and result in a suboptimal total cost or too high a price, but discussing it too late may waste time in discussions about technical details with unrealistically high-priced suppliers.

- When to sign a nondisclosure agreement. Failing to have a supplier sign a nondisclosure agreement at the right time could result in giving away information to competitors, as that supplier bids on work for other companies.

E-business can simplify and speed both contracting and payment cycle times, and additionally may help to articulate and clarify versions of the same product. For example, cataloguing may specify a downhole tool that can operate at 200 meters' depth as opposed to one that can operate at 300 meters' depth—a difference that might not be apparent except through an RFP. It can also easily make distinctions between solution bundles (e.g., the difference between a piece of equipment delivered and installed and the same equipment delivered, installed, and calibrated). Many facets of purchasing and supply chain management can and should be electronic, including

- Vendor preregistration
- Tendering
- Part and service cataloging
- Invoicing
- Ordering
- Auctions
- Contract writing
- Contract administration and compliance monitoring

No matter to what extent the firm has made the tendering process electronic, process simplification can save time and reduce cost. For example, capital and operating purchasing can be done together, if coordinated, and multiple contract types can sometimes be reduced and simplified.

Tendering of products under development before commercialization

Traditionally, major capital projects follow a well-established tendering model. The owner releases information about the project, and potential bidders expresses their interest. The owner then qualifies the bidders and issues an RFP to this short list.

These suppliers then develop a proposal including a firm price and commercial terms. The owner evaluates these bids, negotiates the final agreement with the leading bidder, and awards the contract.

For a project that requires a partnership between the owner and the contractor, the short list of bidders is typically narrowed down to two (instead of three or more) owing to the need to disclose more information about the project, as well as the smaller pool of qualified bidders. Optimally, the project can be defined well enough that the bidders can estimate the total cost to execute it. When this is the case, the owner can follow a *full-price selection* model. Under this model, the owner solicits a full proposal from both short-listed bidders and selects the one with the best value for money after negotiations.[28]

Some projects must be negotiated before the technology is fully commercialized but there is more than one supplier with a solution in the design phase. For example, three original equipment manufacturers (OEMs) have subsea compressor models in the test phase, to compete with only one commercialized model from a fourth OEM. Projects currently in the planning phase that include a subsea compressor therefore conform to a *partial-price selection* model, as the full cost of the project cannot be determined until the compressor design is finalized. Under this model, the owner selects a supplier on the basis of bids for the parts of the project that can be estimated, as well as nonprice criteria such as the test results of the designs under development. Owners must be careful when using this model, as suppliers can underbid on the priced elements of the proposal, knowing that they can make up any difference on the elements without a price tag.

In some cases, a compelling reason exists to forgo a competitive tender entirely. For instance, there might be only one supplier with the necessary technology. In the case of subsea compressors, for example, until 2012 only MAN has had a commercial model. In such cases, the owner must proceed carefully, placing the burden on the supplier to demonstrate that it can execute the project effectively and postponing the actual contract award until a target price has been agreed.

Combined purchase and operating/maintenance agreements

Most OEMs offer to maintain the equipment for a period of time after purchase, and sometimes require it. Sometimes warranty coverage is linked to the OEM maintenance contract. This introduces an economic decision to the operator: OEM maintenance is usually more expensive than in-house or third-party maintenance, but that cost needs to be weighed against the loss of warranty coverage and balanced by higher reliability or uptime if the equipment is maintained by the OEM, rather than a third-party shop.

The market of third-party providers has evolved to encompass a broad range of services such as parts delivery, component reconditioning, spare parts management, rotor repair, testing, remote monitoring & diagnostics, upgrades and conversions, field service, training, inventory management, warranty management, commissioning and startup, and even full operating and maintenance agreements.

The driving factors behind most buyers' decision to use a third party are cost and lead time. A maintenance agreement from a third party is often 25%–50% less expensive than one from the OEM, and third parties can often deliver faster. However, buyers must consider several factors before opting to use a third-party repair firm:

- Depending on the manufacturer and the equipment in question, allowing an unlicensed technician to repair the equipment may void any manufacturer warranties. Moreover, third parties may not be qualified to decouple the unit from the system in which it operates and recommission it when the repair is complete.

- Quality and reliability of the components are also a concern, as third parties have to reverse engineer the parts and reproduce them without the original drawings and equipment. For example, a third party attempting to recreate a gas turbine blade must scan the inside of the blade (using expensive computed-axial-tomography equipment) and go through a trial-and-error process to recreate the serpentine cooling pattern during the casting process.

- Third parties can take up to 10 years to gain enough experience repairing a new model to service the entire unit effectively—especially for complex equipment, such as gas turbines.

- Banks are less likely to finance new equipment purchases, unless the units are under a long-term service agreement (LTSA) with the OEM and/or until service specialists have proved that they can fix the machines effectively.

If the above concerns are addressed, however, then using third-party repair suppliers, who are also sometimes component suppliers, can provide a practical alternative to an OEM service contract.

When looking for a third-party supplier, buyers should look two tiers down in the supply chain—in other words, the suppliers to the OEMs. As demand growth has outstripped production capacity, OEMs have outsourced production of an increasing number of components. This has caused them to delegate a significant amount of control over product quality and delivery schedules to component suppliers. These component manufacturers may, in some cases, provide an effective alternative for service agreements relating to their area of expertise.

Local content management

Operators doing business in foreign countries often need to observe local content laws, as a result of an increasing trend toward *resource nationalism*—countries with mineral wealth are making a concerted effort to channel the wealth generated by the exports into sustainable domestic economic development. Many countries—such as Angola, Brazil, China, and Venezuela—have specific laws requiring local content (table 2–1). In several of these cases, the laws have been enacted since 2001.

Table 2–1. Evolution and status of local content regulations in six countries

Country	Year First Enacted	Scope of Transactions Affected	Obligations under the Laws	Assistance and Facilitation Available
Angola	2003	Oil industry Information technology logistics activities Education sector Industrial hygiene	Hire Angolan staff. Source Angolan products.	Business Support Center (CAE) provides vocational training to help IOCs meet employment requirements.
Azerbaijan	2002	Production-sharing agreements (PSAs)	Contractual agreements come from PSAs rather than legislation. Onshore operations are under joint venture. Offshore operations work under PSAs.	2002 Enterprise Center supports local enterprises win contracts with IOCs.
Brazil	2003	Oil production equipment	Starting in 1997, the ANP used bidder's commitment to local content as a condition for awarding rights to exploration and production. 18% tax on imported oil production equipment (2003) Monitoring and reporting local content (ANP Ordinance 180, 2003) Gradually increasing local content requirements (could go as high as 95% by 2017)	Large contracts may be broken into separate pieces to satisfy local content requirements.

Country	Year	Mechanism	Provisions	Notes
China	2002[a]	Joint venture with Chinese enterprise[b]	Secrecy (Chinese local content requirements are kept in "Neibu" directives) makes precise obligations unclear, inspiring wrath from WTO and EU.	Most energy companies in China use local supplies and labor anyway because of lower cost.
Equatorial New Guinea	2004	Production-sharing agreements; Local supplies and labor	Government right to 20% share in contracts with foreign operators; 2004 forum on local content recommended transfer of technologies to state companies and a new National Institute of Technology.; 2006 hydrocarbon law puts small emphasis on developing local content requirements.; Production-sharing agreement terms emphasize using local supplies and labor.; 13% minimum royalty on petroleum projects	Marathon has started a craft training program to help meet local content requirements.
Venezuela	2001	Exploration and production	2001 Organic Law of Hydrocarbons guarantees exploration and production rights to the state.; Exploration and production must be carried out by state companies or joint ventures in which the state has a stake of more than 50%.	None (foreign companies must develop local content to do business)

[a] marked the end of official local content laws [b] Except for wind turbines, since 2010

Local supply may be more expensive than conventional supply for many reasons, when considering the fixed cost to onboard any new supplier and the possible need for upfront training and investment if the available local suppliers do not meet existing quality standards. The challenges can be diverse and significant—for example,

- Limited technical expertise and supply base because of a lack of professionally qualified labor. In particular, this can apply to managerial talent for precision manufacturing and engineering for capital projects.

- A different level of quality expectation based on local supply market habits, customs, and norms.

- Insufficient information technology infrastructure, possibly including poor phone and Internet access, inadequate, restrictions on content or Voice over Internet Protocol (VoIP), weak Internet security, and outdated or incompatible systems or software.

- Trade restrictions.

- Intellectual property loss owing to reverse engineering or unenforceable contracts.

- Late payment.

- Antiquated logistics infrastructure. Developing countries have worse logistics infrastructure than developed ones, according to the logistics performance index (LPI), a survey-based index established and tracked by the World Bank. Aggregate scores for infrastructure, customs clearance, international shipments, general logistics competence, tracking and tracing, and timeliness of shipments are far lower in developing countries than in mature economies. For example, Nigeria has an LPI of 2.6 and Russia has an LPI of 2.6, compared to 3.9 and 4.1 in the United States and Germany, respectively.[29]

Combining the *hard risks* (e.g., logistics infrastructure) with *soft risks* (e.g., loss of intellectual property) highlights that many countries that are rich in energy resources and/or are involved in major capital investment in the energy industry involve higher supply chain risks than in other countries without those resources (fig. 2–9).

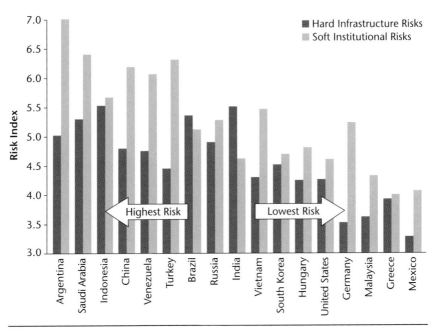

Fig. 2–9. Supply chain risks for 17 countries (*Source:* Jacoby, David. 2011. Uncovering economic and supply chain success in the new emerging economies. Paper presented to APICS International Conference in Pittsburgh, October 24, p. 10).

One drilling equipment manufacturer's experience complying with local content restrictions in West Africa is instructive. This manufacturer needed to employ nine local workers for every expatriate at manufacturing or distribution operations in Nigeria and Angola, with little or no grace period to train the local employees. The official minimum requirement was 70% of the labor force as of 2010,[30] but suppliers needed to maintain this ratio at all times; therefore, they typically stayed well above the minimum level in case local workers quit or they need to bring in more expatriate workers on short notice. Combined with a high turnover rate among the local workforce, the net result of this requirement was that, for every productive worker, foreign suppliers often had several unproductive workers, either actively in training or simply on the payroll, to satisfy local content restrictions.

Other companies have had major successes in the same countries, however, demonstrating that these challenges are manageable. Chevron Venezuela has 80% of goods and 98% of services provided by local suppliers.[31]

FMC bills its Pazflor project as a local content success story.[32] Pazflor is located in deepwater Angola block 17, which is 150 kilometers (93 miles) off the coast of Angola in 600–1,200 meters (1,968–3,907 ft) of water. The development extends over 600 square kilometers (232 square miles) and is operated by Total E&P Angola, a wholly owned subsidiary of Total, with a 40% interest. Statoil (23%), ExxonMobil (20%), and BP (17%) also have ownership stakes. The Pazflor contract was FMC's largest ever, awarded in January 2008, requiring expansion of local production and support capabilities, including installation services and life-of-field support, and extensive local content. FMC delivered 49 subsea trees (25 production, 22 water-injection, and 2 gas-injection trees) and wellhead systems, along with associated production manifold systems, a production control and umbilical distribution system, gas export and flow-line connection systems, and remote operated vehicle (ROV) tooling.[33]

Starting in 2003, Angola required local content for projects in the oil, information technology, education and training, logistics, and hygiene industries. The government's goal is to make all of these sectors Angolan in the sense of being integral to the domestic economy, because of local content required in these industries. These sectors are what the Angolan government has targeted as essential to economic development. Therefore, foreign companies doing business in Angola are required to hire Angolan staff and source Angolan products. The government offers a Business Support Center (CAE) to provide vocational training to help IOCs meet employment requirements.

Local content is also affecting supply chains in Brazil (fig. 2–10). Brazil's National Agency of Petroleum, Natural Gas, and Biofuels (ANP) put incentives in place for local content in 1999–2002. It subsequently made the incentives into requirements in 2003–2004. The percentage of local content required increased from 15.5% in 1999–2002; rose again, to 40%, in 2003–4; and then fell to 20% in 2005–8. In 2005, the ANP imposed the use of certificates of compliance, which defined local content as the total value less anything not produced in Brazil. Rented items are valued by considering their book value, useful life, and the length of time they are active. Suppliers are required to use a certified auditor. In accordance with the law, suppliers must send their credentials to the concessionaire, which must provide the ANP with all suppliers' certificates of local content before it will approve payment. If the targeted thresholds are not reached, ANP assesses penalties.[34]

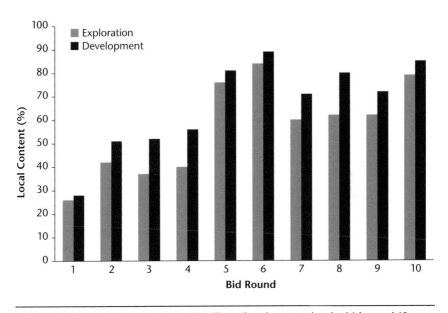

Fig. 2–10. Percent local content in Brazilian oil and gas tenders by bid round (*Source:* Agência Nacional do Petróleo, 2011.)

Clearly, companies wishing to do business in these countries are under pressure to identify and develop high-potential local suppliers. Just as clearly, it is in the government's best interest to support local assistance and training to help route international suppliers to qualified local suppliers. Local content hubs, such as OGPnetwork, have prospered by helping both buyers and suppliers to locate qualified suppliers in oil-rich emerging nations with local content requirements.

When to use auctions

There are six types of auctions: *English, Dutch, Vickrey, sealed bid, combinatorial,* and *buyer-driven.* An English auction, the most frequent and usually the most familiar, is where the item goes to the highest bidder. In Dutch auctions, the price rises in fixed increments, and the first to bid gets the available lot(s) at that price. In Vickrey auctions, the next-to-last bid wins, to avoid a one-penny-more pattern. Sealed-bid auctions are the classic RFP method, with the caveat that there is no negotiation stage. Combinatorial auctions ask all bidders for prices on an array of comparative bundles, and then select the bundle(s) that minimize cost for the buyer. Buyer-driven auctions are essentially English auctions, except that the bidders lack visibility of each other's bids.

The most common auction for business procurement is a reverse auction, in which the suppliers bid successively lower prices until the clock runs out. Reverse auctions have saved 15%–30% for many companies across a range of items where all the specifications are clear and familiar to all the bidders. For example, the U.S. military saved 30% compared to its prebid estimates for telecommunication cable assemblies in a reverse auction.[35] A major international oil company (IOC) saved 16% on the purchase price for one category. Indian energy firm Oil and Natural Gas Corporation saved 12% auctioning its requirements for 100 km of medium-voltage power cable.[36]

Use of auctions has lagged, especially in the oil, gas, and power industries. Reverse auctions have had above-average results in industries such as telecommunications. However, even in the telecommunications industry, less than 15% of companies use reverse auctions for procurement of materials and services.[37] Less than about 5% of all companies currently use auctions, and the oil and gas industry uses them less than most other industries owing to the prevalence of high-specification, safety-critical, high-opportunity-cost applications that prioritize performance before cost.

Despite many efforts to advance the state of the art, auctions do not work well for complex equipment and solutions. Negotiations and multiround bidding can greatly enhance the quality and customization of the solution and adapt it to the application, and this adaptation is often worth much more than can be saved on upfront acquisition cost in an auction. Therefore, oil, gas, and power companies use auctions mostly for MRO procurement today.

Notes

1. As determined by the Herfindahl-Hirschman index: The supply market concentration index is the sum of the squares of the market shares of the suppliers in each category.

2. Froeb, Luke M., and Lance Brannman. 2006. Mergers, cartels, set-asides, and bidding preferences in asymmetric oral auctions. *The Review of Economics and Statistics*. 82. Pp. 283; Ingraham, Allan T. 2005.

3. A test for collusion between a bidder and an auctioneer in sealed-bid auctions. *The B.E. Journal of Economic Analysis & Policy*. P. 9. http://www.bepress.com/cgi/viewcontent.cgi?article=1448&context=bejeap.

4. U.S. Bureau of the Census. EC0731SR2: Manufacturing: Subject Series: Concentration Ratios: Share of Value of Shipments Accounted for by the 4, 8, 20, and 50 Largest Companies for Industries: 2007. 2007 Economic Census. http://factfinder2.census.gov/faces/tableservices/jsf/pages/productview. xhtml?pid=ECN_2007_US_31SR12&prodType=table).

5. Harrison, Jeffery L., and Roger D. Blair 2010. *Monopsony in Law and Economics*. Cambridge, UK: Cambridge University Press. P. 60.

6. Boston Strategies International (formerly Boston Logistics Group). 2005. *Precision-Guided Sourcing Strategies for Maximum Results, High-Impact Sourcing*. White paper. Boston Strategies. http://www.bostonstrategies.com/images/BSI_-_SS2_High-Impact_Sourcing.pdf.

7. Jacoby, David. 2009. *Guide to Supply Chain Management*. New York: Bloomberg Press. P. 64.

8. Froeb and Brannman, 2006, pp. 283–90.

9. *Business Wire*. 2007. Ketera teams with Chevron and Aberdeen Group to host live supplier enablement Webinar. *Business Wire*. May 24. http://www.businesswire.com/news/home/20070524005278/en/Ketera-Teams-Chevron-Aberdeen-Group-Host-Live (accessed May 19, 2012).

10. Zawada, Craig C., Eric V. Roegner, and Michael V. Marn. 2004. *The Price Advantage*. Hoboken, NJ: John Wiley & Sons. P. 258, Appendix 2.

11. Ibid.

12. Stride, Megan. 2011. US sparked global antitrust enforcement in 2010. *Law 360*. January 1. http://www.law360.com/articles/216998/us-sparked-global-antitrust-enforcement-in-2010.

13. PR Newswire. April 27, 2010. Justice department requires divestitures in Baker Hughes' merger with BJ Services. PR Newswire. http://www.justice.gov/atr/public/press_releases/2010/258157.htm.

14. Based on 2008 labor costs in Shanghai, Tianjin, Beijing, and Hangzhou. Boston Strategies International. 2008. China sourcing—the long view.

15. Boston Strategies International. 2006. The Asian sourcing boom: How long will it last? State of Strategies Sourcing. http://www.bostonstrategies.com/images/BSI_-_SS3_The_Asian_Sourcing_Boom.pdf.

16. Best Practices: Human Resources: Strategies for Recruitment, Retention, and Compensation." October 2006. https://www.uschina.org/info/chops/2006/hr-best-practices.html. Last accessed June 3, 2012.

17 Jacoby, David. 2009. The new era of post-Chinese sourcing. *Logistics Digest*. June. P. 22.

18 S Korea's STX Engine inks biz alliance with Rolls Royce section. 2009. *Business in Asia Today*. July 21. http://www.antaranews.com/en/ news/1248167221/s-koreas-stx-engine-inks-biz-alliance-with-rolls-royce.

19 Huntsman. 2010. Huntsman to purchase chemicals business of Laffans Petrochemicals Ltd. Huntsman. http://www.huntsman.com/eng/News/News/ Huntsman_to_Purchase_Chemicals_Business_of_Laffans_Petrochemicals_Ltd/ index.cfm?PageID=8583&News_ID=7949&style=72 (accessed May 19, 2012); *Cosmetics & Toiletries*. 2011. Huntsman Corp. acquires chemical business of Laffans Petrochemicals. *Cosmetics & Toiletries*. http://www. cosmeticsandtoiletries.com/networking/news/company/119403064.html (accessed May 19, 2012); Nalco. [n.d.] Nalco opens new corporate office and technical support center in India. Nalco Web site. http://www.nalco.com/ news-and-events/4657.htm (accessed May 19, 2012).

20 GE Energy. 2010. GE signs licensing agreement for oil & gas compressors with BHEL of India. Press release, July 21. http://site.ge-energy.com/businesses/ ge_oilandgas/en/about/press/en/2010_press/072110.htm (accessed May 19, 2012).

21 Triveni. 2011. Triveni Engineering and Industries Ltd. corporate presentation—August 2011. Triveni Group. http://www.trivenigroup.com/ download/corporate-presentation-aug-2011.pdf (accessed May 19, 2012).

22 LianDi Clean Technology. 2011. LianDi Clean Technology Inc. signs new distribution contract," Press release, January 20. http://www.china-liandi. com/2011January20.html (accessed May 19, 2012).

23 General Electric. 2010. GE and ZKMK sign oil and gas equipment services agreement to support Kazakhstan and Central Asian growth. Press release, July 26. http://www.genewscenter.com/Press-Releases/GE-and-ZKMK-Sign-Oil-and-Gas-Equipment-Services-Agreement-to-Support-Kazakhstan-and-Central-Asian-Growth-29aa.aspx (accessed May 19, 2012).

24 Weir Oil & Gas. 2010. Weir to form China JV focused on shale gas. *Weir Oil & Gas News*. November 24. http://www.weiroilandgas.com/news/weir_oil__gas_ news/2010/weir_oil__gas_to_form_jv.aspx (accessed May 19, 2012).

25 MHI. 2011. MHI Compressor technology licensee in China begins marketing. Press release, February 9. http://www.mhi.co.jp/en/news/story/1102091407. html (accessed May 19, 2012).

26 Bladt Industries to supply Dong Energy with wind turbine foundations. 2011. *Nordic Business Report*. June 28.

27 Sharp, Tim. 2010. SSE buys 15% stake in BiFab. *Herald Scotland*. April 12. P. 1. http://www.heraldscotland.com/business/corporate-sme/sse-buys-15-stake-in-bifab-1.1020234 (accessed May 19, 2012).

28 For further detail, reference the "National Alliance Contracting Guidelines," Commonwealth of Australia 2011, published by Department of Infrastructure and Transport, available at http://www.infrastructure.gov.au/infrastructure/ nacg/index.aspx.

29 Jacoby, David. 2011. Uncovering economic and supply chain success in the new emerging economies. Paper presented to the APICS International Conference, Pittsburgh, October 24; World Bank. [n.d.] Logistics performance index data. http://info.worldbank.org/etools/tradesurvey/mode1b.asp.

30 Nigerian Oil and Gas Industry Content Development Act. 2010. Explanatory memorandum.

31 Menas Associates. Venezuela—local content in practice: encouraging entreupreneurialism with big IDEAS. *Local Content Online.* http://www.menas.co.uk/localcontent/home.aspx?country=61&tab=practice (accessed May 19, 2012).

32 FMC Technologies. [n.d.] Total Pazflor. FMC Technologies. http://www.fmctechnologies.com/en/SubseaSystems/GlobalProjects/Africa/Angola/TotalPazflor.aspx (accessed May 19, 2012).

33 FMC Technologies Web site; Olsen, Willy. 2010. Maximizing the value of strategic partnerships: Introduction—setting the scene. Paper presented to National Oil Congress, London, June 22, 2010. P. 63.

34 Agência Nacional do Petróleo. 2011. Local content framework, the Brazilian experience. Rio de Janeiro, February 6. Powerpoint document by Marcelo Mafra Borges de Macedo. http://www.google.com/url?sa=t&rct=j&q=&esrc=s&frm=1&source=web&cd=1&ved=0CEsQFjAA&url=http%3A%2F%2Fwww.britcham.com.br%2Fdownload%2F020611rj_marcelo_mafra.pdf&ei=2JnLT_vVEui60QGCnZmrAQ&usg=AFQjCNGpLYeGHK7-d9krOqmtaPMayVjEvA.

35 Hopkins, Cathy. 2011. DLA Aviation saves money with reverse auctions. DLA Aviation Public Affairs, May 31. http://www.aviation.dla.mil/externalnews/news/20110531.htm.

36 BOB Tech Solutions. [n.d.] A case study on reverse auction, purchase of signal cables for oil & gas projects development. http://www.google.com/url?sa=t&rct=j&q=&esrc=s&frm=1&source=web&cd=1&ved=0CFcQFjAA&url=http%3A%2F%2Fwww.bobeprocure.com%2Fdocs%2FSignalCables.pdf&ei=kgPMT9WaNqiD0QHnu_GpAQ&usg=AFQjCNEuV6d2ze7QrhXcUJX9Tg6y67FdQQ

37 Checketts, Vance. 2006. *The Cost of Not Acting: The Total Telecom Cost Management Benchmark Report.* Boston: Aberdeen Group, November. P. i.

3

OPERATING AND MAINTENANCE COST REDUCTION: PRINCIPLES AND METHODS

Maintaining Complex Systems

Operating and maintenance costs, including energy, often constitute well over half of total life-cycle cost (with capital expenditure making up the rest), and reliability can be greatly increased by the employment of predictive maintenance using information technology and sensors. Yet planners often estimate operating costs as a flat percentage of capital costs and do not fully leverage the power of best practices in operations and maintenance to drive down cost.

Overall equipment effectiveness

A high-level approach to look at operating and maintenance is by assessing overall equipment effectiveness (OEE). The OEE framework measures asset effectiveness by defining three types of capacity:

- *Rated capacity* (as determined by the OEM).

- *Standard capacity* (the operator's normative, or expected, output). This is driven by equipment availability, which is based on scheduled uptime versus total available time.

- *Demonstrated capacity* (actual production vs. the standard), which is affected by product quality or yield (good output vs. total output).

Actual capacity is the product of rated capacity, standard uptime, and efficiency; or,

$$\text{Capacity} = \text{time available} \times \text{utilization} \times \text{efficiency}$$

where utilization = actual hours worked ÷ standard hours available, and efficiency = standard hours produced ÷ actual hours worked.

The best performers have an OEE averaging 90%, whereas laggards have an OEE averaging 74% according to a study by Aberdeen Group, the differences being largely explained by 2% unscheduled asset downtime versus 12% unscheduled asset downtime. The leaders also experienced 12% reduction in maintenance cost versus 2% increase in maintenance cost, and 24% improvement in ROA versus plan, compared to a 5% decrease in ROA versus plan.[1]

Operational effectiveness depends on cultivating an organization and operational processes and systems that are focused on best practices in maintenance and asset management. Better-performing companies implement preventive and predictive maintenance processes and measure ROA at both the operational and the executive level. Consistent with total productive maintenance, the maintenance and production teams are aligned toward the same goals and therefore jointly contribute to brainstorming and problem-solving. Better-performing companies also spend more time and effort implementing maintenance management systems and asset management systems that tie into their corporate objectives.

Management should set a maintenance strategy either explicitly or implicitly and clarify whether it applies to the whole company or whether it varies by plant, business, or facility. The strategy should correspond to the time horizon of expected active life of the plant. For plants and facilities that are expected to be productive for a long time, the best strategy is generally to maintain the assets to maximize their production for the long run. Occasionally, when a major planned teardown, renovation, or major overhaul is impending, management decides to deliberately let assets deteriorate or to operate until failure.

Even when pursuing an uptime-focused maintenance strategy, it may make sense to repair individual equipment only if it breaks down (e.g., when the cost of taking it off-line for preventive maintenance is not worth the lost output during the downtime when the cost of repairing it would be the same whether it breaks or not). However, most often a preventive maintenance strategy best extends equipment life and results in the lowest long-term cost. There are three basic maintenance strategies, which are often combined into a hybrid solution, especially in large and complex facilities:

- *Periodic*, or *time interval based*, is the simplest and oldest form of preventive maintenance.

- *Usage-based* maintenance schedules visits according to the number of operating hours or starts since the last maintenance, and often lowers maintenance cost if the proper interval can be determined accurately enough.

- *Predictive* maintenance checks for early warning signs of deterioration and takes corrective action before failure occurs. Unlike preventive maintenance, it is not periodic. *Condition monitoring* is a form of predictive maintenance (e.g., the condition of bearings in large machinery can be monitored through electronic sensors and tracked via remote computer) and usually indicates problems well before they occur, facilitating the purchase of replacement parts ahead of time. Remote condition monitoring also reduces on-site maintenance staff requirements and facilitates smoother maintenance workload planning.

Total productive maintenance involves employees in a long-term program of asset productivity optimization through workforce involvement (e.g., quality circles) and the use of problem-solving tools such as root-cause analysis. Reliability-centered maintenance regroups an extensive body of knowledge about maintenance and focuses the enterprise's mission on improving performance toward an overall goal, such as OEE.

Six Sigma, while not exclusively a maintenance concept, can be applied to maintenance. Six Sigma is the application of statistical process control to reduce variation in performance levels. It seeks conformity to a target level, which results in deviations only in the rarest of occasions. Six standard deviations from the norm represent only 3.4 defects per million items, which is equivalent to a result that is correct in 99.9997% of occurrences.

Total cost of ownership

Many configurations of supply relationship involve upfront costs and ongoing operating and/or maintenance costs. On the basis of total cost of ownership (TCO), companies can evaluate complex equipment in terms of lower-cost ownership. Although the concept is simple, few companies have a reliable framework for calculating TCO, and many decisions are made based on gut feeling, historical precedent, or political expediency.

TCO, also referred to as life-cycle cost, represents the total cost of acquiring, commissioning, operating, maintaining, and disposing of a product or system (fig. 3–1). For the technologically complex and frequently expensive equipment sourced by an oil and gas or power producer, this equation includes substantial costs, before and after the initial purchase, that dwarf the acquisition cost alone. Before a piece of equipment is bought, it must be specified and suppliers selected, a process that, for heavy equipment such as injection pumps, often requires a study. The equipment must then be transported to the point of use, which can involve multiple modes of transport, handling insurance, and customs duties. Installation and inventory costs also need to be considered in TCO, including such hidden factors as property taxes, warranty expenses, and storage costs. Throughout the operating life of most electrical equipment, maintenance and energy costs eventually outweigh the initial purchase price. For example, a 500-cubic-foot-per-minute oil field air compressor with an initial cost of $24,000 will use about $120,000 in electricity over a 10-year period and will cost $16,000 to maintain.[2] Finally, decommissioning costs can be substantial, particularly for offshore platforms with subsea structures that must be dismantled and removed.

A comprehensive measurement of the cost of ownership should include 10 elements:

- Forecasting, including engineering resources and studies that would not have otherwise been required

- Acquisition, including the cost of the equipment itself and any ancillary equipment required to make it operational; if the equipment is being repurposed, the acquisition cost would consist of the cost of converting or adapting equipment from a previous use

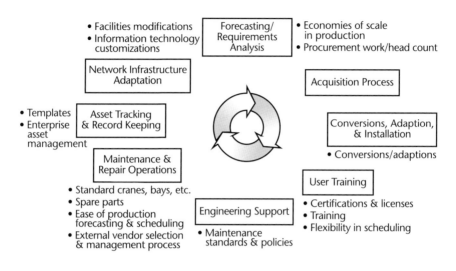

Fig. 3–1. Components of TCO (life-cycle cost)

- Commissioning, including testing and calibration
- User training and engineering support
- Operation, including power, man-hours of any operating personnel, and allocated facility costs
- Maintenance, including maintenance staff and facility cost and testing or measurement equipment
- Inventory of spare and repair parts
- Asset tracking and record keeping
- Infrastructure adaptation to suit the configuration, including software and networking costs related to the unit
- End-of-life removal and disposal, including environmentally safe disposal costs

Quality costs and related concepts

Some people count quality costs as part of TCO. In many cases, however, since the cost of quality failure does not affect the equipment itself, it is accounted for separately so that the life-cycle costs can be used to compare and evaluate alternative equipment choices, whereas quality costs can be used to improve internal processes and systems. Quality costs include the following:

- Internal failure costs primarily comprise the cost of rejects, reworking, reinspection, and replacement. However, additional internal failure costs include corrective action, wasted material or other inputs, late charges, expediting costs, and early costs (i.e., opportunity costs).

- External failure costs include returns, warranty claims, liability costs (penalties/allowances), loss of customer goodwill, and lost sales.

- Prevention costs include identifying customer needs (market research), reviewing contracts, testing new products, certifying suppliers, and the cost of running a quality department and processes (e.g., statistical process control).

- Appraisal costs include source inspection, incoming inspection, the cost of measurement and laboratory equipment and services, and evaluation of field stock or service quality.[3]

Zero defects is a quality philosophy that emphasizes that ensuring good quality through conformance to requirements is cheaper than inspection and rework.[4] Cause-and-effect analysis is a useful tool for engaging worker involvement in solving complex operating- and maintenance-related problems.[5]

Constraints management, debottlenecking, and flexible capacity

The capacity of a system is limited to the capacity of its bottleneck process. Therefore, capacity should be aligned at each step to maximize the capacity of a supply chain. Consider the case of an LNG export supply chain that involves liquefaction, tanker shipping, and regasification on the other side. If the heat exchanger used in the liquefaction process has a higher capacity than the rate at which the tanker can berth and load, then the heat exchanger will be underutilized. Similarly, if the capacity of the storage tanks or the seawater warming on the delivery side is lower than the capacity of the tanker or of the liquefaction process, the upstream process capacity will be underutilized. Moreover, in the latter case there will be a bottleneck.

Debottlenecking is the process of successively identifying the binding constraint, eliminating that constraint, aligning other processes to the new throughput levels, and pursuing the next constraint. Debottlenecking efforts should consider the entire supply chain as a system. This is challenging because, to identify bottlenecks, supply chain partners would need to share information about their processes and many companies hesitate to share such information as it could be considered as confidential business information.

As demand levels change over time, capacity must flex up and down. Given that capacity takes a long time to build and remove in oil, gas, and power, there have historically been two basic strategies for matching supply to demand: *chasing* and *leading*. Chasing means adjusting capacity in *response* to changing demand, while leading means adjusting capacity in *anticipation* of changing demand. Both approaches assume some risk—in the case of chasing, the risk that capacity will not be available when needed and, in the case of leading, the risk that capacity will sit idle until demand catches up. These risks can be minimized through *flexible capacity management*. Flexible capacity management lowers the breakeven point, thereby increasing profit potential without increasing the asset base.

During ongoing operations (post-capex), five tactics can be used to make capacity flexible:

- Increase temporary employees as a proportion of staff. This reduces the fixed labor cost base and therefore lowers the breakeven point. Temporary employees are easier to hire and let go when demand shifts and if managed carefully can eliminate expensive overtime.

- Implement *lean* concepts—such as just-in-time, one-piece flow, level loading, cycle-time compression, and make-to-order—to avoid building unnecessary inventory.

- Postpone disposition, dispatching, or finishing operations to as late as possible in the value chain in order to reduce capacity requirements and inventory obsolescence.

- Use demand-planning techniques—such as sales and operations planning and collaborative planning, forecasting, and replenishment—to reduce overstock and out-of-stock situations.

- Dynamically price products, such as lubricants in retail outlets, to encourage customers to buy stock before it becomes obsolete—and, conversely, to earn extra margin on tight resources during peak periods. Once customers adapt their buying behavior to yield pricing, this has the effect of leveling capacity.[6]

Standardization

Standardization of items and services—the reduction of the number of types or brands of equipment or the number of component variations—can increase the purchased unit volume per item, thereby lowering the average purchase price, as well as the life-cycle cost of spare parts, maintenance, repair, and training. The case for standardization usually rests on the costs of not standardizing. For example, by operating multiple different types or brands of electrical equipment, an operator would need multiple spare-parts pools, each possibly with its own service network. There would be overlapping and potentially conflicting documentation and training efforts, resulting in a duplication of effort and redundancy of resources. In turn, the redundancy would eventually cause some confusion, which would increase the likelihood of errors and deviations to operating procedures, which would reduce reliability and uptime.

In contrast, the potential benefits and savings from standardization are speed (e.g., faster time to first oil and to start-up of a new plant based on the use of familiar equipment as well as standardized facility design and construction), quicker installation, lower upfront acquisition cost based on volume discounts, lower training cost, lower operating and maintenance cost, and lower financial (e.g., technology and warranty) risk. Standardization programs are often best directed at operating processes and systems by limiting the number of suppliers, technology platforms, or component types, rather than at the actual configuration of equipment (which could reduce the suitability for diverse applications and operating environments, thereby reducing reliability).

Process standardization includes harmonization of work processes and operating procedures, work instruction cards (for maintenance), and design engineering practices. Most major oil companies have implemented standard process initiatives, of which the following are examples:

- Shell has standardized design and engineering practices (DEP), which list the specifications and uses of various pieces of equipment. Combined with its Materials and Equipment Standards and Code (MESC) catalog, Shell has cut inventory levels in half for electrical cable and has achieved 30% price savings.[7]

- ExxonMobil actively participates in determination of standards with the International Organization for Standardization (ISO) and the American Petroleum Institute (API) and has performed case studies on specific standards in support of standardization.[8]

- Petrobras launched its Propoço program to standardize well construction. The program established standards for projects, personnel, well design, and documentation. It encourages the sharing of information and best practices and offers engineers detailed studies, calculations, and designs to optimize well construction.[9]

- Gazprom created a corporate standardization system in 2002, consisting of technical guidelines for product requirement specifications, and standardized processes in all aspects of the business. In addition to enabling cost savings, standardization is expected to increase conformance to governmental and internal regulations.[10]

- Petronas compiled preapproved technical practices as a technical reference for engineers and a guideline for vendors and contractors.[11]

ERP platforms are a good place to begin standardizing systems. Many companies have multiple instances (versions, upgrades, modules, etc.) of the same ERP system in place, providing a high-level standardization opportunity. In addition, data definitions, data governance, and data formats and protocols are all rich targets for standardization initiatives. Petro-Canada has supported common standards in the industry; with a presentation at the International

Standardization Workshop in Doha in 2006, it advanced common standards as a way to share best practices, delocalize procurement and investment, and reduce cost and delivery times.[12]

Standardization bodies, whether at the international or the national level, facilitate standardization. Examples include the International Organization for Standardization (ISO) and International Electrotechnical Commission (IEC), the American National Standards Institute (ANSI), the British Standards Institution (BSI), NORSOK (Standards Norway), OLF (The Norwegian Oil Industry Association), and regulatory bodies from Brazil, Japan, and European countries. Organizations like CEN (Comité Européen de Normalisation) and CENELEC (CEN's electrotechnical division) bridge international and national standardization bodies. Various associations composed of contractors, suppliers, and operators from many industries complement the efforts of these standards bodies. For example, API and ASME (American Society of Mechanical Engineers) write standards that are reviewed and accredited by ANSI, whereas EEMUA (Engineering Equipment and Materials Users Association) and IP (The Institute of Petroleum) are British organizations that write standards administered by BSI. In these cases, the official standards bodies audit the standard-writing procedures of the industry groups and recertify them periodically.

Over time, the standards of the different regional, industry, segment, and engineering disciplines are converging. For example, the Norwegian oil industry has developed 80 harmonized specifications, called *NORSOK standards*, freely available on the Internet. The effort started in 1993 to make drilling on the Norwegian Continental Shelf more cost-competitive by replacing company specifications and detailed regulatory requirements to cut cost by up to 50% and reduce project completion time by up to 25%. The standardization has saved money by shortening project schedules, eliminating redundant engineering, and reducing both capital and operating costs.[13] NORSOK had a goal to replace NORSOK standards with international standards as soon as possible. This effort has resulted in fewer company specifications, and the Norwegian Petroleum Directorate (NPD) has reduced the volume of its regulations from over 1,200 pages to just 300 pages through the use of referenced standards, including both international and global standards such as those of API, IEC, ISO, and NACE.

Achieving Continuous Cost Reduction

Continuous cost reduction can drive opex costs down by 1%–5% through initiatives such as flexible demand planning, proper management of erratic inventory consumption, consignment and vendor-managed inventory programs, logistics outsourcing (i.e., third- and fourth-party logistics service providers), and lean distribution, including just-in-time.

Spare-parts inventory forecasting

For basic inventory management requirements, oil, gas, and power inventory management uses the same tools as other industries—demand forecasting methods, MRP and reordering or replenishment methods, stockkeeping rules, and so forth. Inventory serves the basic purpose of satisfying demand during the order lead time; this is called *cycle stock* and is normally constant across multiple cycles of production and ordering.

Spare parts are considered to be independent demand items because they are drawn directly according to an end-user requirement, rather than being a component in another item that is being manufactured. Independent demand items are most frequently replenished using reorder points (whereas dependent demand items are replenished using bill-of-material explosion and offsetting techniques in MRP systems). The reorder point is calculated as follows:

Reorder point = demand during lead time + safety stock

where demand during lead time is the number of weeks times units per week and safety stock is the standard deviation of demand times a safety factor. Reorder quantities are often determined by calculation of the economic order quantity (EOQ). (Other replenishment methods include *min-max*, *total cost*, and *grid search*.)

The challenge that the oil, gas, and power industries face—as do all process industries, from paper milling to aluminum, glass, and cement production—is that some items, particularly capital spare parts, have an irregular demand pattern that is especially hard to forecast, and a single out-of-stock incident, or *stock-out*, of even a small item could shut down production. Intermittent demand makes inventory planning difficult and forecasts unreliable. As a result, materials managers often keep a buffer of just-in-case inventory (safety stock), which inevitably doesn't prevent all stock-outs. Sooner or later, financial controllers blame inventory managers for what looks like excess inventory in some items and simultaneous high-visibility stock-outs—surely somebody must not be doing their job!

Spare parts are different from other inventory problems in five ways, adapted here to oil, gas, and power applications:[14]

- They have unpredictable demand that is stochastic and intermittent. Because of this, the replenishment rule is usually to lot-for-lot or one-for-one owing to the high value and customized applications of each unit.

- The target stock level is often zero (order on demand) or one (one-for-one reorder), which increases the chance of a stock-out considerably compared to a high-volume, small-value item that is stocked by the hundreds. Quite frequently, supplier lot sizes are larger than EOQs, so the reorder quantities are larger than the reorder point, which means that there is *overstock* as soon as each replenishment order is received.

- MRP systems, if set to replenish according to the classic algorithms, draw the wrong conclusion from an item consumed at irregular frequencies. They classify it as volatile and thereby replenish more than required. Over time the replenishment ends up being driven more by exaggerated restocking patterns than by actual consumption.

- The value of replacement parts, especially critical spares, is high and widely variable (inventory carrying costs for capital spares comprise 25%–35% of total inventory value), and stock-out costs can be disproportionate to the cost of acquiring or stocking the unit.

- Stock-out costs are hard to calculate and are disputable when they are calculated. If a one-for-one replenishment of a high-torque drive belt for a drawworks on an oil rig needs to be replaced and the rig will be shut down if the belt is not available, then the stock-out cost can be hundreds of thousands of dollars per day. However, most managers will not use such numbers in the computerized maintenance management system as the basis for inventory calculations, since the calculation will most likely conclude that the reorder quantity level should be inordinately high.

- Maintenance and life-cycle history records may need to be maintained on each part moving through inventory, which can be time consuming, be detail oriented, and require system integration of a records management system separate from the maintenance management information system. In the aircraft industry, some parts must be scrapped if the complete history of the part since its manufacture cannot be produced, due to the risk of counterfeit or other safety-related problems that may have occurred during periods that are unaccounted for.[15]

The best way to determine spare-parts requirements with stochastic demand is by statistical failure analysis on the equipment itself rather than by looking at inventory consumption patterns. A number of specialist software applications apply probability to historical reliability and failure data—and sometimes to location-specific inventory storerooms—to ascertain how many spare parts should be purchased and where they should be placed to best respond to potential requirements. This is especially true for capital spare parts, such as turbine rotors or air compressor screws, which can cost half as much as the compressor itself.

Other ways to reduce capital spares requirements include:

- Outsourcing maintenance, transferring responsibility for spare parts to the service company—which frequently is the manufacturer. For instance, pump suppliers Sulzer and Flowserve have well-publicized maintenance programs that feature uptime guarantees for critical equipment.

- Standardizing equipment so that the same spare part can be used in multiple pieces of equipment. This method is most effective when capital equipment is concentrated in a small geographic area, such as for refinery expansions or multistage oil field developments.

- Reducing the lead time for a new spare part. For example, if a rotating equipment operator sources a spare gearbox from a supplier that can deliver in three weeks instead of six, then the operator can reduce its pool of spare parts.

Flexible inventory management

Mechanisms for *flexible inventory planning* allow a flexible response to changes in demand and outlook. Methods of flexible inventory planning include, through dynamic inventory management, monitoring the downstream demand and increasing the frequency of demand, as follows:

- Managing each class and volatility combination uniquely and dynamically. If fast-moving inventory is Class A and slow-moving inventory is Class B and if volatile demand items are Type 1 and stable demand items are Type 2, then for A1 inventory replenish on a just-in-time basis and engage in financial hedging and for B1 inventory add an extra layer of safety stock.

- Tracking the demand of customers' customers, rather than that of immediate customers only. The added *anticipation lead time* increases the chances of having adequate time for replenishment or of reducing inventory rapidly in response to a sharp drop in demand.

- Increasing the frequency of demand (production, sales, and inventory) planning so that fewer items are unavailable as a result of being caught between replenishment cycles.[16]

Vendor-managed inventory

As Slater has aptly said, there are only two ways to reduce inventory: to take items out of inventory and to put less into inventory. Vendor-managed inventory (VMI) is one way to take items out of inventory, by requiring a supplier to be responsible for them.[17]

For discrete product delivery into the oil, gas, and power supply chain, traditional models of consignment and VMI work just fine. So long as there are no hazardous goods, chemicals that change characteristics with pressure and temperature, or other peculiarities, consignment (paying the vendor when the item is used rather than when it is purchased) and VMI (having the vendor determine the stocking level and holding him accountable for the fill rate) work the same as in any other industry.

One example of VMI is New Pig's Flawless Logistics and Optimized Warehousing system to reduce lead times and inventory holding costs. This system includes warehouse management, transportation network optimization, and customer relationship management software to improve customer service and speed the resolution of any complaints. By implementing the new system, New Pig reduced response time to initial inquiries from three days to less than 24 hours and increased on-time delivery rates from 95% in 2009 to 99% in 2011.[18] Another example is Baker Hughes' VMI system for oil and gas production chemicals at the well site, which eliminates the need for operators to set feed rates, check tank levels, and replenish chemicals. Chemicals managed under the program include corrosion inhibitors, sweetening agents (to remove hydrogen sulfide [H_2S]), biocides, defoamers, clarifiers, and scale inhibitors. Baker Hughes' wireless tank-monitoring network, SentryNet, determines the right feed rate for a given chemical on the basis of composition and volume of the production stream. SentryNet also keeps track of the amount of chemicals in a tank and sends an alert to Baker Hughes when that tank needs to be refilled or when the rate of chemical addition differs from the preprogrammed level (indicating a leak or a plug). Baker Hughes then dispatches a truck to refill the tank or fix the equipment. This often replaces a manual monitoring system on the part of the operator, which requires a technician to follow a physical inspection route every few days to check tank levels and replenish them if necessary.

Operators have complete visibility to the process, as they have access to the same data stream Baker Hughes receives, including tank levels, fill dates, and repair logs. Thus, Baker Hughes can bundle the requirements of several operators in an area and optimize efficiency by serving the entire area, instead of just one customer. As evidence of the effectiveness of the program, Baker Hughes has pointed to the operations of a customer in the U.S. Haynesville shale, where it has managed more than 100 wells without a corrosion-related failure since 2010.[19] Baker Hughes also uses the system to send automatic alerts to its own chemical suppliers, according to preset reorder points, alerting them to incoming orders when the tank at the well site is depleted, instead of waiting until the chemicals ship from their own facility.

Nalco also upgraded its chemical delivery service, adding a 400-gallon drum size to its previous range of 30–200-gallon drums. Under this service plan, Nalco monitors customer inventory levels remotely, delivers new chemical drums to the customer's location, and manages spent drums, reducing inventory for the operator.[20]

Just-in-time inventory management

As mentioned above, the second way to remove inventory is to put in fewer items. One of the ways to do this is to replenish only what has been consumed, which is the essence of just-in-time inventory management.

Just-in-time stems from lean thinking at Toyota in the 1950s. Taiichi Ohno, vice president of manufacturing, and Shigeo Shingo, head of industrial engineering and factory improvement training, developed the Toyota Production System, emphasizing the minimizing all waste and doing it right the first time. Important elements of the Toyota Production System are Kanban, uniform plant loading, reduced setup times, and workforce involvement.

Just-in-time production systems rely on the use of physical inventory control cues (Kanban) to signal the need to move or produce new raw materials or components from the previous process (fig. 3–2). It often requires suppliers to deliver components using just-in-time inventory management. The buyer signals its suppliers, using computers or delivery of empty containers, to supply more of a particular component when they are needed. This results in significant reduction in waste associated with unnecessary inventory, work-in-process (WIP), and overproduction. Removing buffer inventory also exposes problems so they may be addressed. As more and more slack is removed from the system, all process inefficiencies (fig. 3–3) become apparent and are removed through continuous improvement. If applied repetitively, this results in no WIP inventory, short lead times, short order-cycle times, and small lots, and in theory, a one-piece flow.

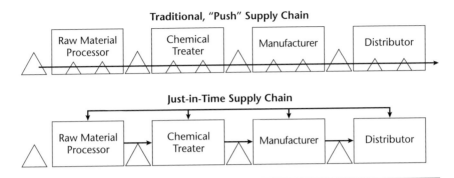

Fig. 3–2. Traditional push versus just-in-time supply chain (*Source:* Jacoby, David and Michael Gourd. 2005. "Lean Distribution: How Distribution is Changing, and Tools You Will Need for the New Environment." Seminar given to Boston APICS, May 5. P. 21).

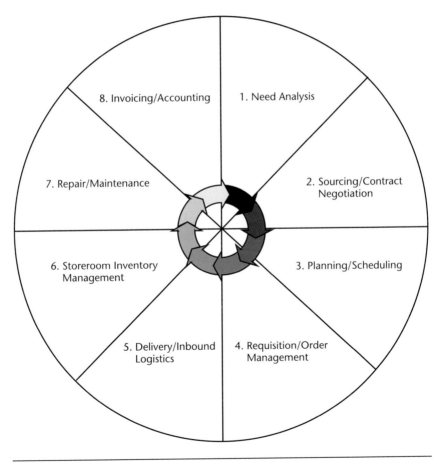

Fig. 3–3. Maintenance, repair, and operating (MRO) order management processes

Inventory is one of the seven deadly wastes identified in lean thinking. (The other six are overproduction, waiting time, transportation of any kind, processing downtime, motion, and waste from product defects, which includes scrap, rework and recalls.) Similarly, just-in-time is one of five core methods of lean. (The other four are Kaizen, 5S, total productive maintenance, and cellular manufacturing.)

Lean is often described as a philosophy, or even a religion. One lean consultant described lean as a set of principles that includes putting the customer first, having an end-to-end total value-stream perspective, focusing on consistent flow, and eliminating waste.[21] The breadth of scope of lean makes it both very important (the reason why it has had such a profound impact on business) and potentially enigmatic (the range of interpretations of its scope can seem to defy boundaries).

Lean can be defined in many ways (while a full background on lean is beyond the scope of this book, interested readers may find several excellent texts dedicated to the subject; see app. D for further reading suggestions). The purpose here is merely to explain how just-in-time fits into oil and gas supply chain management. The following tools have yielded exceptional results in waste elimination programs:

- The seven types of waste
- The 5S approach to workplace organization
- Total quality management
- Total productive maintenance
- Cellular manufacturing
- Diagnostic tools, such as the five whys, the plan-do-check-act cycle, value-stream mapping, work sampling, root-cause analysis, and throughput analysis.

Lean implementations have realized tremendous benefits. One study of 80 plants in Europe found that they had reduced inventory by about 50% and throughput time by 50%–70%. The plants cut setup times by similar proportions without major investment in plant and equipment and increased productivity by 20%–50%. Moreover, the payback time for the investment in just-in-time management averaged less than nine months.[22]

The benefits of lean are well recognized within the oil and gas industry. For example, Chevron won Boston Strategies International's 2010 supply chain award for effectively extending its internal lean Six Sigma techniques to its suppliers. Examples of improvement projects include the streamlined truck routing for KS Industries (a piping installer), affecting more than 200 employees, and the standardized vehicle fleet for Braun Electric, which installs Chevron's electrical systems for oil field pumps. The initiative has reduced average drilling time, workover rig cycle time, and lead time for engineering drawings.[23]

Since achievement of such benefits almost always requires the joint participation of supply chain partners, it is common to *gain-share*, or share the benefits of Lean with partners. Some companies split the benefits 50/50, whereas others have designed asymmetric approaches. One asymmetric approach allocates the benefits to the buyer in early periods and then fixes a minimum expected amount above which the savings will be shared with the supplier.[24]

Transportation and warehousing optimization

Improved transportation routing and scheduling can offer a quick way to reduce operating costs, especially if the traffic or trade management functions are currently handled manually or if there is variation in service performance either internally or externally. Variations of logistics cost reduction initiatives include the following:

- *Automation.* Transportation management systems reduce the amount of manual rate checking and booking tasks.

- *Mode optimization.* Cost and service can frequently be optimized when expedited air shipments are shifted to ground transport or change from one mode to another.

- *Cross-docking.* A subset of shipments can sometimes be transloaded at an intermediate point without being stored, as part of a rapid-delivery hub-and-spoke network.

- *Distribution center bypass.* Moving freight directly to its end destination without storing it at a distribution or even a mixing center can reduce cost and transit time.

- *Equipment pooling.* Sharing fleet assets, usually through a third-party equipment-pooling company, can reduce per-mile cost. Pooling is often used for railcars and intermodal chassis.

There are numerous success stories within each of these categories, including the following:

- The Middle East division of a major oil field services company reduced customs clearance time, increased transit time reliability and visibility to cargo status, provided customers multiple cost versus delivery options, and reduced shipping costs by switching its distribution center hub location and transport mode for imports from Europe and the United States.

- A major IOC proved that mode selection can be applied creatively. It developed a high-volume crude oil rail shuttle to transport crude oil from an important shale play to refineries 1,500 miles away, thereby increasing the yield from the region, which was constrained by a lack of outbound transportation capacity, and simultaneously increasing the efficiency and operating margins of the refinery by increasing its utilization rate.

- A North American steel company shifted tubular volume from truck to rail, saving nearly $1 million in annual freight costs for shipments to Canadian oil fields.

- One pipe distributor saved $300,000 per year by transferring its tubular freight volume from a truck carrier to an industrial freight broker. Whereas the truck carrier had charged round-trip rates since it ran empty on backhauls, the industrial freight broker had enough volume to fill backhauls and charged only one-way rates.[25]

Total logistics outsourcing

When companies hire third-party logistics companies (3PLs), they often gain capabilities, flexibility, and responsiveness. Therefore, they should not base the outsourcing decision on cost savings alone.

Use of 3PLs can save money on operating costs, but the savings is a decreasing part of the equation. Because of their economies of scale, 3PLs can often purchase packaging, supplies, fuel, and equipment at lower unit prices, then pass these lower costs on to their customers, thereby reducing their operating costs. Historically, 3PLs have rehired union employees into nonunion environments, resulting in an immediate cost savings, although sometimes with encumbrances as employment obligations lapse or as the union employees quit or retire. Unionization rates have declined dramatically in the past two decades (from 20% in 1983 to 12% in 2007 in the United States),[26] though, consequently reducing this incentive to use 3PLs; moreover, unionization is even less of an issue in developing economies such as Asia's.

While operating cost savings is not always the driving force, outsourcing logistics can be financially advantageous insofar as it transfers assets to the 3PL. Off-loading tangible assets, such as warehouse facilities and vehicles, and intangible assets, such as information systems, increases ROA. Higher ROAs make companies more attractive to lenders and shareholders, making it easier to obtain loans and float shares. Furthermore, a one-time improvement through outsourcing can result in an uptick in stock price.

A significant benefit of hiring a 3PL is the capability to serve customers better. The following strategic advantages favor hiring a 3PL over doing the job in-house:

- First, 3PLs often have a more global collection and distribution network than many shippers. Their global scale can allow shippers to extend their sourcing and distribution beyond the scope of their existing networks. The sourcing advantage can provide greater cost benefits than shippers may have initially hoped to capture by outsourcing logistics. The distribution benefits may allow them to position for sales in emerging geographies that would have been impossible to reach.

- Second, 3PLs invest heavily in information technology platforms that provide measurably higher service levels and service reliability compared to in-house logistics operations. Best-of-breed transportation management systems and warehouse management systems provide the ability to handle complex pick/pack operations more readily than in-house solutions.

- Third, 3PLs invest in emerging technologies (e.g., radiofrequency identification [RFID] tags) sooner than their customers do because they can amortize the investment across multiple customers. This results in earlier adoption of important information technology capabilities than would have occurred in-house.

- Fourth, 3PLs also offer value-added services, such as custom packaging operations. Examples include clamshell packaging, security packaging, kitting, customized packaging by customer, or personalized packaging by consumer (e.g., through individualized insertions).

The logistics outsourcing and selection process should be made in alignment with the overall business strategy. Used correctly, logistics outsourcing can be a key element of effective supply chain management. Therefore, the decision should be made at the senior executive level, and decision-makers should integrate the decision with the company's supply chain management strategy.

Petroleum Development Oman (PDO) successfully outsourced its entire logistics function (i.e., fourth-party logistics [4PL]) to BE/DHL. In 2005, PDO signed a 4PL contract with Bahwan Exel/ DHL, a consortium of Bahwan Cybertek and DHL Supply Chain.[27] The initial years of the contract were focused on rig moves, but the group eventually took over complete oversight of transport management processes and systems at PDO.[28] The association led to 10%–20% improvements in trip planning and journey management and simplified operations through a centralized organization structure. In addition to cost savings, the partnership brought technological benefits, for example, Web-enabled booking requests, cargo tracking and tracing, more accurate paperwork, and more meaningful management reporting.[29]

When logistics outsourcing agreements fail, it is often because of poor communication and unclear or misaligned goals. Outsourcing arrangements should be based on broad solutions that affect the bottom line in order to avoid unproductive behaviors such as overemphasis on cost reduction at the expense of good service; expecting or requiring the 3PL to replicate suboptimal in-house procedures; pay per unit of activity, which motivates activity rather than cost savings; and insufficient input or collaboration from the buyer.[30]

Notes

1 Ismail, Nuris. 2011. *Enterprise Asset Management in 2012*. Boston: Aberdeen Group.

2 Airpower UK. [n.d.] Energy savings using variable speed drive air compressor technology. Airpower UK. http://www.airpowered.co.uk/article1.htm (accessed May 19, 2012).

3 Juran, Joseph. 2010. *The Complete Guide to Performance Excellence*. New York: McGraw-Hill.

4 Crosby, Philip B. 1995. *Philip Crosby's Reflections on Quality: 295 Inspirations from the World's Foremost Quality Guru*. New York: McGraw-Hill.

5 Ishikawa, Kaoru. 1991. *What Is Total Quality Control? The Japanese Way*. Englewood Cliffs, NJ: Prentice-Hall.

6 Jacoby, David. 2010. Using flexible capacity techniques to thrive in a volatile economy. *Logistics Digest*. February 9. http://www.logisticsdigest.com/inter-education/inter-opinion/item/4669-using-flexible-capacity-techniques-to-thrive-in-a-volatile-economy.html (accessed May 19, 2012).

7 Boston Strategies International. 2010. 2010 Oil and Gas Award for Excellence in Supply Chain Management awarded to Shell. Press release, September 22.

8 ExxonMobil. [n.d.] Profile: Angola (brochure).

9 *Drilling Contractor*. 2008. Petrobras' ambitious PROPOÇO a "road map" for optimizing well construction performance. *Drilling Contractor*. January/February. P. 138.

10 Gazprom. 2009. Management committee approves Gazprom's standardization and technical regulation efforts. Press release, April 1.

11 Pau, K. H. 2009. Utilization of PETRONAS Technical Standard. Presentation to Petronas Research and Technology Division, February 23. http://www.google.com/url?sa=t&rct=j&q=utilization%20of%20petronas%20technical%20standard%20(pts)&source=web&cd=1&ved=0CCYQFjAA&url=http%3A%2F%2Finfo.ogp.org.uk%2Fstandards%2F09Malaysia%2FPresentations%2F11SharingOnPTSFeb09v10.pdf&ei=K584T56RLYPq0gHi3KyZAg&usg=AFQjCNFIsyiNSfon00tnhGMf2vfff3RIQ (accessed May 19, 2012).

12 Lever, Gregory. 2006. Participation of the Americas in international standardization effort and the role of international industry: regional and national organizations and need for standards. International Standardization Workshop, Doha, Qatar -- 3 April 2006. International Standardization Workshop. Presentation, Doha, Qatar, April 3.

13 American Petroleum Institute. [n.d.] The oil and natural gas industry's most valuable resource. American Petroleum Institute Web site. http://www.api.org/Standards/faq/upload/valueofstandards.pdf (accessed May 19, 2012).

14 Phillip Slater authored an easy-to-read reference on spares inventory management, to which credit is due for some of the points in this section and the subsequent two sections. See Appendix D for bibliographical details.

15 Adapted from Slater, Phillip. 2010. *Smart Inventory Solutions: Improving the Management of Engineering Materials and Spare Parts*. New York: Industrial Press.

16 Murphy, Bruce. 2010. How has the recession changed the way we plan? Paper presented to Boston APICS, Waltham, Massachusetts, February 23. P. 12.

17 Slater, 2010.

18 New Pig. [n.d.] Why New Pig? Orders go out the door within 24 business hours. New Pig Web site. http://www.newpig.com/us/content/why-new-pig?title=Why%2520New%2520Pig%3F (accessed May 19, 2012).

19 Baker Hughes. 2011. Upstream chemical service plan delivers annual savings of more than $2 million. *TechConnect*. 9 (3). http://enewsletter.bakerhughes.com/techconnect/_vol9no3_landing.htm (accessed May 19, 2012).

20 New Pig Web site.

21 Carreira, Bill. 2005. *Lean Manufacturing That Works: Powerful Tools for Dramatically Reducing Waste and Maximizing Profits*. New York: American Management Asssociation. P. 3.

22 Sohal, Amrik, and Keith Howard. 1987. Trends in materials management. *International Journal of Production Distribution and Materials Management*. 17 (5). Pp. 3–11.

23 Boston Strategies International. 2010. Chevron wins Boston Strategies International's 2010 Award for Lean Six Sigma Implementation in Oil and Gas Operations. Press release, September 22.

24 Underhill, Tim. 1996. *Strategic Alliances: Managing the Supply Chain*. Tulsa, OK: PennWell. P. 110.

25 PLS Logistics. 2012. Freight transportation in the oil & gas industry: five mistakes that cripple profitability. White paper.

26 Schmitt, John, and Ben Zipperer. 2007. Union rates fall in 2006, severe drop in manufacturing. Center for Economic Policy and Research. January 25. http://www.cepr.net/index.php/union-membership-bytes/union-rates-fall-severe-drop-in-manufacturing-in-2006/.

27 Kennaugh, Rob. 2009. 4PL from a provider's perspective: Building and implementing a 4PL—the journey is tough; the reward is mutual. Paper presented to TransOman, Muscat, Oman, October 14.

28 Rice, Stephen. 2009. 4PL from a client's perspective. Paper presented to TransOman, Muscat, Oman, October 14.

29 Kharusi, Warith. 2009. Creating logistics synergies within Oman. Paper presented to TransOman, Muscat, Oman, October 14.

30 Geary, Steve. 2009. Performance-based outsourcing: The next generation of outsourcing. Paper presented to Boston APICS, Waltham, Massachusetts, May 19.

4

OPERATIONAL SAFETY AND ENVIRONMENTAL RISK MANAGEMENT: PRINCIPLES AND METHODS

At the extreme, operational risk can wipe out project profitability (e.g., by cutting off repatriation of profit). More often, though, it is only one of a number of soft risks that threaten efficient execution of project plans. Operational risks can be classified into nine categories:

- Logistics infrastructure
- Regulatory
- Economic dependence
- Market growth
- Legal institutions
- Unavailability of skilled labor
- Closed market
- Supply
- Intellectual property

When quantified, six of the risks threaten project execution and profitability more than the others: labor availability, competitive playing field, information technology infrastructure, trade restrictions, intellectual property security, and logistics infrastructure (fig. 4–1).

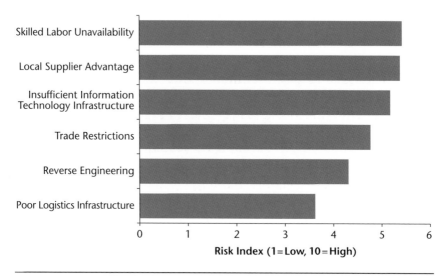

Fig. 4–1. Top supply chain risks, based on an analysis of 14 emerging countries, 2011 (*Source:* Jacoby, David. 2011. Uncovering economic and supply chain success in the new emerging economies. Paper presented to APICS International Conference. Pittsburgh. October 24. P. 11.)

These risks need to be managed at every step of the capex project process—including engineering, procurement, construction, operations, maintenance, and decommissioning. Furthermore, they need to respect principles of transparency and honesty with supply chain partners. This chapter provides frameworks and tools for managing supply chain risk. The information is divided among the following sections:

- Risk management standards, primarily at the international but also at the national and corporate levels
- Operational risk mitigation frameworks
- Environmental footprint mitigation protocols (i.e., supply chain's role in reducing environmental footprint, including carbon footprint)

Selected International Risk Management Standards

Agencies such as ISO, the International Maritime Organization (IMO), and the United Kingdom's Office of Government Commerce (OGC) have developed risk management frameworks. For example, since the Macondo Field oil spill, ISO has developed a new action plan (ISO/TC 67) that pertains to materials, equipment, and offshore structures for the petroleum, petrochemical, and natural gas industries. The action plan provides an inventory of 71 existing standards and related documents from ISO and other organizations, particularly API. It also proposes 31 standards or related documents for development or update by these same organizations. The Directorate General of the European Union has also proposed new rules to incorporate technological improvements. It proposes removing minimum technical requirements and replacing them with a goal-setting approach that will assess risk on a case-by-case basis and determine the best risk-mitigation approach on the basis of the best technology currently available.

Several important risk management standards are discussed below.

ISO 31000

ISO 31000, like the AS/NZS 4360 standard upon which it was built, contains generic guidelines that any organization in any context and scope can use for effective risk management. It considers risk management as an integral part of all organizational processes and an effective part of decision-making. The principle of ISO 31000 is that risk management must be structured, timely, adapted to each organization's needs, and transparent to all in the organization. Insofar as management should define and endorse a risk management policy that is in line with the company's objectives and legal and regulatory compliance, designing a framework for risk management must involve an understanding of the internal and external context of the organization, including its internal and external communication and reporting mechanisms.

ISO 31000 also describes attributes of high-level performance in managing risks and lists indicators for organizations to compare their performance against these criteria. Key indicators for enhanced risk management include the following:

- A programmatic approach emphasizing continuous improvement in risk management through an explicit performance goals and review process

- Articulated and explicitly accepted accountability for risk management and mitigation embedded in job descriptions

- Inclusion of risk management in meeting agendas, and active record keeping of discussions about risk.

- Communication about risks with internal and external stakeholders

ISO 31000 provides guidance for formulating a policy on risk, offering, for example, the following elements:

- Governance structure for risk management and internal control, including assignment of roles and responsibilities, and dedication of sufficient resources toward risk management

- Explicit statement of the risk strategy of the firm

- Profile and description of the culture as it relates to risk

- Procedures and supporting documentation for identifying risks and prioritizing them for action

- Statement of appropriate responses to all types of risk

- Risk management training

- Annual risk management plan[1]

ISO 31010 also provides a large number of tools to quantify the consequences, probability, and level of risk, as outlined in table 4–1.

The government of Alberta announced plans to implement ISO 31000 to "develop and implement a risk management approach to regulating upstream oil and gas projects."[2] Qatar Petroleum initiated a program to train its senior management in risk management in October 2011.[3] The mining firm TWP Projects has also adopted an ISO 31000 enterprise risk management (ERM) system.[4]

Table 4–1. Summary of risk management tools (*Source:* ISO 2008, p. 24. Permission to reproduce extracts from BS 31100:2008 is granted by BSI. British Standards can be obtained in PDF or hard copy formats from the BSI online shop: www.bsigroup.com/Shop or by contacting BSI Customer Services for hardcopies only: Tel: +44 (0)20 8996 9001, Email: cservices@bsigroup.com.)

Tool	Identification	Assessment	Response
Risk questionnaires	×		
Risk checklists/prompt lists	×		
Risk identification workshop	×	×	
Nominal group technique	×	×	
Risk breakdown structure	×	×	
Delphi technique	×	×	
Process mapping	×	×	
Cause-and-effect diagrams	×	×	
Risk mapping/risk profiling	×	×	
Risk indicators	×		
Brainstorming/"thought shower" events	×		
Interviews/focus groups	×		
What-if workshops	×		
Scenario analysis/scenario planning/horizon scanning	×	×	×
Hazard & operability study	×	×	
PEST analysis[a]	×	×	
SWOT analysis[b]	×	×	
Stakeholder engagement/matrices	×		
Risk registry/database	×	×	×
Project profile model	×		
Risk taxonomy	×		
Gap analysis: Pareto analysis	×	×	
Probability & consequence grid/diagrams/Boston grid	×	×	
Central Computer and Telecommunications Agency Risk Analysis and Management Method	×	×	×
Probability trees		×	
Expected-value method		×	
Risk modeling/simulation[c]		×	
Flow charts, process maps & documentation		×	
Fault and event tree modeling[d]		×	
Stress testing	×	×	
Critical-path analysis/method		×	
Sensitivity analysis		×	
Cash-flow analysis		×	
Portfolio analysis		×	
Cost-benefit analysis		×	×
Utility theory		×	
Visualization techniques[e]		×	×

[a] Political, economic, sociological & technological. [b] Strengths, weaknesses, opportunities & threats. [c] Using Monte Carlo/Latin hypercube methods. [d] Using failure mode/effects analysis. [e] Including heat maps; red, amber, and green status reports; waterfall charts; profile graphs; 3D graphs; radar chart; and scatter diagram.

Committee of Sponsoring Organizations integrated framework

The Committee of Sponsoring Organizations (COSO) of the Treadway Commission developed a framework for ERM in organizations, which came to be known as the COSO *integrated framework*. COSO defines ERM in terms of the strategy that an organization follows to manage risk within the firm's *risk appetite* as it strives to achieve its objectives. The COSO framework identifies risk management as an iterative process—as opposed to a sequential process, where each step has a bearing on the others. It aims to augment the internal control structure within an organization for effective ERM.

The COSO framework is built upon six objectives:

1. Aligning risk tolerance with the business strategy
2. Encouraging risk avoidance and risk-sharing
3. Reducing unexpected incidents as well as financial and nonfinancial losses
4. Identifying and managing risks across multiple departments and business units in a coordinated and consistent way
5. Bias for action based on outstanding risks
6. Deploying capital to where it can most reduce risk[5]

The framework describes the following eight components of ERM (paraphrased here):

1. Setting the tone for risk management, including corporate culture, ethical norms, and values that underpin operational decisions
2. Articulating objectives related to risk
3. Identifying risk-prone situations
4. Assessing risks
5. Formulating explicit risk management strategies and tactics that align with corporate goals
6. Establishing procedures to ensure that risk management activities are performed as planned

7. Dissemination of information that enables people to manage risks the way the organization plans for them

8. Overall management of the ERM process[6]

Formal safety assessment (IMO)

Formal safety assessment (FSA), originally developed in response to the Piper Alpha disaster of 1988 when an offshore platform exploded in the North Sea, taking 167 lives, is a systematic process for assessing the risks associated with shipping activity. FSA consists of five main steps:

- "Identification of hazards (list of all relevant accident scenarios with causes and damages)

- Assessment of risks (evaluation of risk factors)

- Risk-control options (regulatory measures to control and reduce the identified risks)

- Cost-benefit assessment (determining cost-effectiveness of each risk-control option)

- Recommendations for decision-making (information about the hazards, their associated risks, and the cost effectiveness of alternative risk-control options is provided)"[7]

Other Risk Management Standards

There are dozens of other risk management standards, mostly at the national level. However, discussion of all of them is beyond the scope of this book.

One standard worth mentioning is the U.K.'s Management of Risk (MOR) framework, which focuses on risk management in public sector organizations. The MOR framework was first published in 2002 by the U.K. Office of Government Commerce, which is a part of the Efficiency and Reform Group of the Cabinet Office, in response to the Turnbull Report (a U.K. study that addressed standards for internal control and risk transparency).

Another notable standard is Det Norske Veritas's Integrated Software Dependent Systems (ISDS). Data security has become a much larger concern since the advent of smart grids. Information system security standards have become more prevalent as smart grid implementation revealed that data breaches could shut down operations and compromise safety. For example, Det Norske Veritas developed the ISDS standard beginning in 2008, and Dolphin Drilling received the first ISDS class notation.[8]

Operational Risk-Mitigation Frameworks

While international standards are intended to support ERM approaches, other frameworks may be more useful for identifying and managing risks at the project level. In particular, companies each take their own approach. For example, Qatar Petroleum has instituted risk-based inspection. The company's Professional Training Division offered a five-day training course on risk-based inspection aimed at maintenance personnel, operations supervisors, and process specialists expected to make decisions regarding the suitability of equipment for continued service. It sought to establish and implement a risk-based inspection program best suited for the company objectives. The course was based on API Recommended Practice 580 and API Publication 581, both on risk-based inspection.[9]

Lukoil implemented risk prevention through operational standard operating procedures. In 2003, Lukoil adopted its Fundamentals of the Strategy and Policy of Insurance Protection. This policy intends to protect operations, employee health, and interests of shareholders. Moreover, it aims to identify industrial risks through surveys and quantitative assessments engage in risk mitigation procedures, including operations changes and additional investment where necessary, and purchase insurance protection.[10]

Also, the Council for Security Cooperation in the Asia Pacific released a memorandum titled "Safety and security of offshore oil and gas installations," in January 2011, which identifies new safety and security risks resulting from an increase in offshore drilling activity. The memorandum urges regional governments to implement safety standards—such as those put forward by the Association of Southeast Asian Nations in the 2005 Agreement on Disaster Management and Emergency Response—and to dialogue with operators to agree on their respective responsibilities and develop joint emergency-response plans.[11]

For immediately adaptable frameworks dealing with risk, managers may want to adapt either the root-cause framework, the probability-versus-severity cube (i.e., classifying the likelihood vs. its severity, then ranking the overall risks on the basis of the combined score[12]), or failure mode/effects analysis. However, the most universal problem-solving framework that could be applied to supply chain risk framework may be a root-cause framework (fig. 4–2). (For a comprehensive risk management toolbox, interested readers may consult the additional sources listed in app. D.)

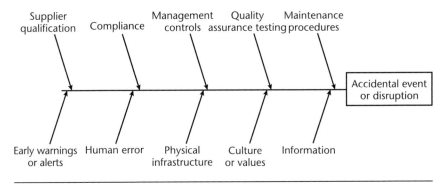

Fig. 4–2. Cause-and-effect analysis applied to supply chain risk

Failure mode/effects analysis

Failure mode/effects analysis is a method to identify and prevent process failure. Planners most effectively use this analytical method during product development and process planning. The exact steps can vary, depending on who describes the process; nevertheless, a generalized and comprehensive description of failure mode/effects analysis would include the following steps:

- Gather a cross-functional team of people who touch the process at the beginning, the middle, and the end

- Define what output the system is intended to produce
- Identify where it can go wrong (failure mode)
- Identify the effects, or consequences, of failure (qualitatively)
- Develop a scoring sheet, framework, or matrix for capturing the group's assessment of severity, frequency, and detection on a common scale (e.g., a spreadsheet with scoring indices from 1 to 10) as it relates to the relevant time frame, such as the lifetime of the equipment or process
- Quantify the severity of the potential failure
- Quantify the frequency of the potential failure
- Quantify the chance of the failure's being detected early enough to be acted upon (using an inverse scale, so a high score represents a low chance of detection)
- Calculate the risk priority number as severity × frequency × chance of detection
- Identify ways to reduce the frequency of failure, to increase the chance of early detection, or to reduce the severity of the failure (the drivers of the risk priority number)
- Establish a tracking log, with responsibilities and dates[13]

Separation of duties

Separation of duties can reduce the possibility for errors that might lead to safety incidents. Valero and Areva both implemented an application for separation of duties. Valero Energy implemented a process for identifying and remediating segregation of duties conflicts to comply with the Sarbanes-Oxley Act by improving access compliance enterprise-wide. It selected access-control software to standardize the process for compliance assurance, as opposed to the fragmented compliance initiatives that had previously been in place. Areva implemented a tool that allows the flexibility to design customized risk matrices for the company's multiple divisions, and leverage preventive controls built into the business processes to stop future violations.

Supply Chain's Role in Reducing Environmental Footprint

Increasing pressure from government regulatory bodies, society, and consumers has led to an increasing interest in measuring environmental footprints and reducing carbon emissions. This initial pressure has been compounded by a natural business desire for efficient processes and awareness that green processes are often more efficient, which has been exacerbated by rising prices of resources that are being depleted and supply shortages of some raw materials.

A broad movement throughout recent history has compelled supply chain managers to monitor and comply with environmental legislation. Below are highlights of this history that should be familiar to a supply chain manager, especially if he or she is working in the United States:

- 1970. The U.S. Environmental Protection Agency (EPA) and Occupational Safety and Health Administration (OSHA) were formed.

- 1976. The Resource Conservation and Recovery Act (RCRA) established laws surrounding the disposal of hazardous waste.

- 1976. The EPA enacted the Toxic Substances Control Act (TSCA).

- 1980. The Comprehensive Environmental Response, Compensation, and Liability Act (CERCLA) was enacted.

- 1986. The Superfund Amendments and Reauthorization Act (SARA) was amended.

- 1986. Right-to-Know laws established hazardous communication standards, such as MSDSs.

- 1992. The United Nations' Basel Convention was entered into force, preventing transfer of hazardous waste from developed to less developed countries.

- 1997. The Kyoto protocol was adopted.

- 2003. The Waste Electrical and Electronic Equipment Directive dealt with the problem of persistent electronic waste.

- 2004. The Stockholm Convention on Persistent Organic Pollutants became effective 2004, restricting the production and use of persistent organic pollutants.

- 2004. The Rotterdam Convention promoted an open exchange of information and calls on exporters of hazardous chemicals to use proper labeling.

- 2006. The International Restriction of Hazardous Substances Directive was instituted.

- 2008. The deadline arrived for preregistration of substances under Registration, Evaluation and Authorization of Chemicals (REACH) (see app. D).

- 2010. The United Nations' Year of Biodiversity and other initiatives encouraged a widespread focus on climate change.

Supply chain's role starts from material procurement and goes through changes in product and process design to final delivery to an end customer. Therefore, ecologically conscientious practices—such as green product design (especially in the case of energy-consuming equipment, e.g., motors, generators, and pumps), green materials management (transport modes, packaging, etc.), and reverse logistics or recycling—can make a significant impact. Companies in other industries have in some cases implemented highly structured models for incentivizing suppliers to be green, and it is likely that some energy producers will emulate this practice in the future.[14]

Environmental risk management monitors and addresses potential environmental breaches. Faced with the need to prove environmental compliance following community complaints, KNPC chose an environmental, health, and safety and crisis management solution program, to respond to changing environmental regulations. Kuwait's government has recognized KNPC's efforts as a best-practice model.[15]

Many are measuring their carbon footprint. The Carbon Disclosure Project (CDP) is an independent not-for-profit organization with a mission of accelerating solutions to climate change and water management by disseminating information to businesses, policy-makers, and investors. Over 3,000 organizations in 60 countries participate by measuring their greenhouse-gas emission, water management, and climate change strategies through CDP, so they can set reduction targets and make performance improvements. These data are made available for use by institutional investors, corporations, policy-makers and their advisors, public sector organizations, government bodies, academics, and the general public.

CDP's Supply Chain and Public Procurement initiatives may be of interest to supply chain professionals working in the oil, gas, and power industries. CDP's Supply Chain initiative works with global corporations to understand the impacts of climate change across the supply chain, harnessing their collective purchasing power to encourage suppliers to measure and disclose climate change information. CDP's Public Procurement initiative is designed to enable national and local governments to ascertain the impact of climate change in their supply chains.

In the area of electricity consumption, demand response has gained a lot of traction in recent years, as has the transition from old, inefficient lighting technologies to new, far more efficient ones (e.g., compact fluorescent bulbs). In addition, efficient motor technologies based on IE2 and IE3 efficiency standards have been implemented. These trends affect the choice of suppliers and technologies, and ultimately the cost and efficiency of production of oil, gas, and power.

In the area of air emissions, new air emission regulations in the United States and European Union for engines will affect prices and availability of equipment, necessitating an equipment transition and new maintenance strategies for end users.[16] The EPA's new air emission standards for spark ignition engines rated less than 500 horsepower will go fully into effect in 2013. The EPA aims to reduce annual toxic emissions by 6,000 tons, particle pollution by 96,000 tons, carbon monoxide (CO) emissions by 109,000 tons, and volatile organic compound emissions by 31,000 tons annually. Engine operators will need to retrofit their engines or replace with them with new model engines, which will cost over a third more than the existing ones. Similarly, the European Union will implement Euro 6 in 2013, which will decrease emissions of particulate matter by 50% and oxides of nitrogen (NO_x) by 75%–80% over the previous (Euro 5) standards.

Oil and gas companies have been responding with plans to reduce carbon dioxide (CO_2) emissions. Saudi Aramco released a carbon management roadmap in 2006, outlining five avenues for reducing carbon emissions (namely CO_2 capture—fixed sources; CO_2 capture—mobile sources; CO_2 EOR; CO_2 sequestration; and developing industrial applications that use CO_2 as a feedstock), both from its operations and from transportation applications that use fossil fuels, such as passenger cars. The firm is a major sponsor of the Weyburn-Midale CO_2 Storage and Monitoring Initiative, a research project run by the IEA to study the feasibility and best techniques for combining EOR with carbon capture sequestration (CCS). Saudi Aramco plans to begin a pilot CO_2 EOR project in 2013.[17]

In the area of chemicals and wastewater, environmental pressures have led to a host of new green biocides, decontamination and neutralization techniques, chemical, electrical and ultraviolet treatments to remove pollutants and undesirable bacteria from produced water.[18] Green efforts intensified after Nalco and BP came under public pressure for extensive use of the dispersant Corexit 9500, which Louisiana residents claimed is four times more toxic than the leaked oil and other viable dispersant options.[19] A number of innovations demonstrate a strong commitment on the part of the industry to improve the environmental impact of chemicals and wastewater—for example,

- NexLube, an independent U.S. start-up, will produce lubricants entirely from recycled waste oil when it opens a plant in Florida.[20]

- Drillers are recycling wastewater. For example, each well in the Marcellus shale consumes 100,000–300,000 gallons of water for drilling (excluding fracturing). One drilling company recycled 80% of its water in 2009, 90% in 2010, and 100% in 2011.[21]

- Thousands of companies are complying with REACH's far-reaching and time-consuming registration process.[22]

ISO 14001 provides a framework for a holistic approach to environmental policy for organizations. Like ISO 31000, it defines generic requirements for an environmental management system as a reference standard for communicating between companies, regulators, and other stakeholders. Draka, the Dutch cable supplier, implemented environmental management systems for all operations in compliance with ISO 14001.[23]

To make a significant impact, supply chain managers need to work with suppliers and customers and with supply chain partners two or more links away in the supply chain. This brings up the trust issues that are more fully addressed in *The Guide to Supply Chain Management*.[24] Supply chain partners are traditionally loathe to share information that may reveal confidential or sensitive business information, especially related to costs and anything that may carry liability, such as carbon footprints or environmental damage caused by their accidents. However, compelling, vocal, and well-informed leaders will work with both near and far supply chain partners to better secure both safety and the environment.

Notes

1 A structured approach to enterprise risk management (ERM) and the requirements of ISO 31000. 2010. The Institute of Risk Management. P. 11. http://theirm.org/ISO31000guide.htm.

2 Alberta to better integrate oil and gas policy and regulatory system. 2011. *Pinoy Times*. February 25. http://pinoytimes.ca/2011/02/alberta-news/alberta-to-better-integrate-oil-and-gas-policy-and-regulatory-system/.

3 Quality Austria Gulf. [n.d.] We Serve Group W.L.L. has successfully been certified accreditated to ISO 9001:2008, ISO 14001:2004 & OSHAS18001:2007. Quality Austria Gulf Web site. http://www.qualityaustriagulf.com/html/newsandevents.html (accessed May 19, 2012)

4 *Miner's Choice*. [n.d.] Heads and tails of risk: Edmond Furter interviews TWP Sherq officer Quinton van Eeden. *Miner's Choice*. http://www.minerschoice.co.za/sheq2.html (accessed May 19, 2012).

5 Enterprise Risk Management—Integrated Framework. 2004. Committee of Sponsoring Organizations of the Treadway Commission. P. 1. http://www.coso.org/guidance.htm.

6 Ibid., pp. 3–4.

7 International Maritime Organization. Formal safety assessment. http://www.imo.org/OurWork/Safety/SafetyTopics/Pages/FormalSafetyAssessment.aspx.

8 Richardsen, Per Wiggo. 2010. New class notation for integrated software dependent systems released. DNV press release, May 18. http://www.dnv.com/press_area/press_releases/2010/newclassnotationforintegratedsoftwaredependentsystemsreleased.asp (accessed May 19, 2012).

9 Qatar Petroleum Risk Based Inspection (RBI) Course. http://www.qp.com.qa/qp.nsf/print.htm (accessed June 9, 2010).

10 Lukoil. [n.d.] Insurance strategy and policy. Lukoil Web site. http://www.lukoil.com/static_6_5id_266_.html (accessed May 19, 2012).

11 CSCAP. 2011. Safety and security of offshore oil and gas installations. CSCAP memorandum no. 16, January.

12 Jacoby, David. 2009. *Guide to Supply Chain Management*. New York: Bloomberg Press, 2009. P. 177.

13 Dumke, Daniel. 2011. Predicting and managing supply chain risks. *SCRM Blog*. November 21. http://scrmblog.com/review/predicting-and-managing-supply-chain-risks (accessed May 19, 2012).

14 Nissan. *Green Purchasing Guide*.

15 IHS. [n.d.] KNPC advances industry best practices by expanding enterprise-level EHS information management. IHS Web site. http://www.ihs.com/ar/images/KNPC-2011_Excellence_Award.pdf (accessed May 19, 2012).

16 U.S. Environmental Protection Agency. [n.d.] Final air toxics standards for reciprocating internal combustion engines. EPA fact sheet. http://www.epa.gov/airtoxics/rice/rice_neshap_fr_fs081010.pdf (accessed May 19, 2012).

17 Al-Meshari, Ali. 2008. Saudi Aramco's carbon capture and sequestration technology road map. Saudi Aramco presentation, January. http://www.cslforum.org/publications/documents/SaudiArabia/T2_2_CSLF_CM_TechRoadMap_Saudi_Aramoc_Jan08.pdf (accessed May 19, 2012).

18 Jacoby, David. 2011. "Global Trade Restrictions and Related Compliance Issues Pertaining to Oil and Gas Production Chemicals." Society of Petroleum Engineers technical paper # OTC 22005-PP, May 2011. Prepared for presentation at the Offshore Technology Conference, Houston, May 2–5.

19 Green production chemistry. 2010. University of Stavanger, Norway, June 8. http://www.uis.no/research/natural_sciences/chemistry_and_environment/oil_field_production_chemicals/ (accessed May 19, 2012).

20 Joyce, Matt. 2012. "Addison's NexLube opening $85M Florida plant." *Dallas Business Journal*, June 29. http://www.bizjournals.com/dallas/print-edition/2012/06/29/addisons-nexlube-opening-85m-florida.html.

21 Hopey, Don. 2011. Gas drillers recycling more water, using fewer chemicals. *Pittsburgh Post-Gazette*. March 29. http://www.post-gazette.com/pg/11060/1128780-503.stm?cmpid=news.xml (accessed May 19, 2012).

22 REACH Web site. http://ec.europa.eu/environment/chemicals/reach/reach_intro.htm (accessed May 17, 2012).

23 Draka. [n.d.] Environmental management. Draka Web site. http://www.draka.com/draka/lang/en/nav/Sustainable_business/Environmental_management/index.jsp (accessed May 19, 2012).

24 Jacoby, 2009.

Part 2

5

UPSTREAM OIL AND GAS EXAMPLES

Introduction: Supply Chain Cost Drivers and Relevant Design Constructs

The extremely engineering-intensive nature of exploration might make it seem that supply chain management is peripheral to upstream oil and gas. The formal training of most personnel in exploration is typically scientific (e.g., geologists, seismologists, stratigraphers, paleontologists, and geophysicists) and is frequently based in petroleum engineering (e.g., reservoir engineers, petrophysical engineers, drilling engineers, and production engineers).[1] Seismology, for example, involves more statistics than logistics. Decision science applications are often technical more than business-oriented. For example, tools such as Crystal Ball and @RISK are used in reservoir modeling to estimate and probabilistically express the reserves that are trapped in formations.[2]

Nevertheless, the rigs themselves are often rented, rather than owned by the oil companies, and their operation is often outsourced, which opens up all the same issues about supply chain that were discussed in Part 1. Furthermore, many of the personnel operating drilling rigs are subcontracted oil field service employees, including the drilling crew (e.g., rig manager, driller, derrickman, roughnecks, and roustabouts), production crew (e.g., production foreman and pumpers/gaugers and roustabouts), and craftspeople (e.g., electricians, mechanics, pipefitters, plumbers, instrument technicians, carpenters, welders, metalworkers, and drivers).[3]

In addition, the work of exploration relies on a specialized supply chain of activities related to consumable products and services, such as drilling fluids (muds), bits, well fracturing and stimulation (including sand control services), formation evaluation (wireline and mud logging, measuring while drilling/logging while drilling, and coring), and well completion (services, liner hangers, packers, perforation, and flow control), service tools (coiled tubing and drillable, retrievable, and stimulation tools), fishing, oil and gas production chemicals and distribution systems, artificial lift, well servicing, contract compression, laboratory services, production chemical services, and water control.

Upstream capital procurement includes compressors (mostly reciprocating for infield gas gathering or storage and injection compressors), high-pressure pumps for injection and use on production platforms, and turbines and generators for production facility power or drive power.

Project Risk Mitigation

The massive capital at risk in upstream oil and gas makes producers and their suppliers especially vulnerable to market risk, operational risk, and financial risk. Because the capital costs are high, most rig operators concentrate regionally to leverage central supply points and their familiarity with the area (e.g., Diamond Offshore in the Gulf of Mexico and Unique Maritime in the United Kingdom).

Noise is created by fluctuations in rig rental activity and flows through the supply chain. These fluctuations, characterized as *bullwhip*, are passed through the supplier capital investment cycle, resulting in oscillating prices, capital investment, and facility utilization. Appendix A contains a detailed study of the long-term effects of this chain reaction. Suppliers experience the bullwhip effect more than producers do, because the bullwhip effect is magnified as it is transmitted through the supply chain (fig. 5–1).

Oil & Gas Production (Oil Rigs)
Cost of supply chain misalignment for oil and gas producers

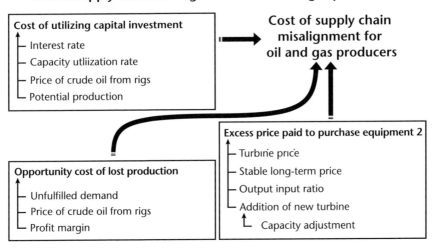

Fig. 5–1. Upstream oil and gas supply chain (*Source:* Adapted from Jacoby, David. 2010. The oil price bullwhip: problem, cost, response. *Oil & Gas Journal.* March 22, p. 21).

Bullwhip causes a vicious cycle of inefficiency. Symptoms of delayed reaction to changes in drilling volume include:

- Rising and falling capacity utilization
- Extending and contracting lead times
- Price inflation and, occasionally, deflation

Suppliers have used vertical integration, scale, and market dominance to shield themselves against the bullwhip effect over time, which explains in significant part why the industry is so concentrated for oil field products and services.

Oil companies have structured operations to mirror incremental and successive levels of financial commitment. The exploration stage has historically been managed incrementally. Planning is extensive and occurs through a multistage vetting process. Forecasting error rates (wells that are not commercially viable) begin high during early exploration, when it is relatively cheap to fail, but decline as operators reach project stages that require more investment. The main stages of project development are:

- Concept studies: 25%–40% error

- Feasibility studies: 10% error; however, these represent 1%–2% of the total project cost

- Front-end engineering and design (FEED) studies: 5% error; these cost about 5%–10% of the total project cost

- EPC: <5% variance to budget[4]

Once facilities are in operation, successive and incremental investments limit exposure to large losses. For example, after geological studies determine that the remaining reserves justify the investment, primary production (natural flow, artificial lift, gas injection, hydraulic lift, and plunger lift) is supplemented by secondary recovery (water flooding and, in a repeating pattern, five-spot flooding of multiple injector wells near a single producing well). Similarly, operators sometimes opt for tertiary recovery (fire floods, steam floods, and CO_2 injection) if the reservoir characteristics justify it.

Financial and real options have become a more common as a way of dealing with risk related to pre-FID commitments. The option to lease land is one of the oldest options tools (fig. 5–2). The land option cost varies, depending on whether the option to lease is coupled with shooting rights (it is expensed if it is not coupled or capitalized if it is coupled) and whether any of the acreage is leased (it becomes capitalized if it is leased or expensed if it is not leased).[5]

Historically, oil and power companies have also sought to share the risk with partners, for example, through production sharing contracts in Iraq (BP), Oman (Transco), United Arab Emirates (Abu Dhabi Water & Electric Authority/GDF Suez), and Qatar (AES-Gulf Investment).[6] In such arrangements, responsibilities are split between the IOC and the NOC. Traditionally, the IOC would fund exploration, appraisal, development, and production and then transfer ownership of assets to the NOC. The NOC would subsequently provide power and water and a share of the profit.[7] In recent times, ever greater risk is being transferred to the service providers. Most LSTK tender guidelines are clear to shift risk to the supplier in the event of liquidated damages for delay and defects liability. They also require IOCs to put up performance bonds and agree to indemnities and liens.

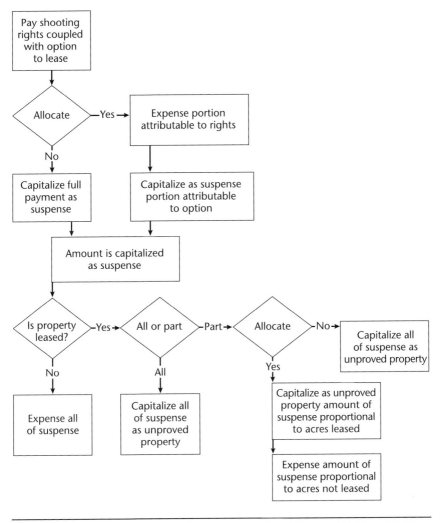

Fig. 5–2. Decision flowchart for taking options on land (*Source:* Wright and Gallun 2008, p. 101.)

Today EPC firms are much less inclined to enter into LSTK contracts, which allocate nearly all risk to the EPC firm. Especially for projects valued at over $500 million, most of these contracts have recently earned less than expected or have lost money owing to escalating labor costs, unrealistic time constraints and penalties imposed by the owner, and owner micromanagement of the process, including stipulations about which technologies and subsuppliers to use.[8] EPC firms increased prices to offset their risk, which has made EPC work less competitive with in-house work and general contracting than in the past. The balance has shifted to the extent that NOCs and most major players in the wind power industry, which is entering a rapid phase of major capital investment, are not looking toward EPC firms at all, preferring traditional methods of construction contracting.

The industry is now using more-sophisticated methods to determine how much risk each party should bear. *Real-options analysis* can help to better hedge against the risk of failure, as well as of raw material price inflation and equipment unavailability. Texaco and Exxon used real options in the 1990s for exploration and production decisions. Mobil used real options in 1996 to decide whether to develop a natural gas field, and, also in the 1990s, Anadarko used them to bid for oil reserves.[9] However, most companies have not used real-options analysis for applications involving supplier risk and market risk. Potential applications include deciding on the optimal contract length or choosing between different supply chain risks that could stem from few manufacturing locations (e.g., decentralized vs. centralized production).

Risk modeling has been used to determine the optimal contract term for certain types of upstream supply contracts. One analysis tested sensitivities on different price discounts (10%, 20%, 30%, or 40%) and found that a buyer who secures a 10% price discount for a 1.5-year contract could secure a 20% price discount for a 5-year contract.[10] The relationship between contract term and price discount is specific to products and services and changes over time with market volatility.

Very-long-term contracts can minimize risk-related costs in oil and gas, according to a study of an oil and gas supply chain (see app. A). A simulation model measured the impact of volatility over 43 years on four types of inefficiencies (high prices, excess inventory, excess capacity investments, and lost orders) at four nodes in the supply chain: refiners, upstream oil producers, equipment OEMs, and component suppliers. The model tracked seven economic variables (backlog, orders, production, prices, capacity, capacity utilization, and total supply chain cost) under a scenario in which the price of oil was flat and under a scenario in which the price of oil fluctuated. The conclusion was that there are two basic strategies for managing supply chain risk in the face of uncertainty: *go short*, continually rebidding contracts and tracking the market (this is a type of hedge); or *go long*. To eradicate the impact of production-inventory-capacity cycles, companies need agreements longer than 20 years. For this to be achieved with minimal risk, it was concluded, capital-intensive companies could form contracts that are long enough to ride through both ups and downs but provide flexibility to share cost risks and leave options for each party to adjust their commitments related to capacity, lead times, and prices as conditions change.

In addition to long-term contracting, operators can assess risk in their capex scenarios by including acquisition cost uncertainty and operating/maintenance cost uncertainty in the NPV calculation. By use of this approach, three figures should be hard numbers— namely, acquisition cost, operating/maintenance cost, and number of units. Acquisition cost risk decreases as a result of learning-curve effects and operating/maintenance cost risk decreases with the time since model-year launch, as a result of a declining rate of infant mortality. Total cost can then be calculated as the sum of: (1) acquisition cost; (2) acquisition cost risk; (3) operating/ maintenance cost; and (4) operating/maintenance cost risk.

Managing Supply Availability and Price Risk

Raw materials and equipment or its major components are prone to becoming bottlenecks, but the supply risk can be managed by passing it through, hedging it, sharing it, and eliminating it (or trying to). For example, lead times for castings have fallen slightly, but castings suppliers have not added capacity and are still backlogged—meaning that castings remain bottleneck items.[11] Similarly, barite has also been in short supply. These are just two of many such examples.

Whenever there is a cost increase, the easiest path is to pass this risk directly and immediately on to customers in the form of a surcharge if the market will bear it. Suppliers can either raise prices by the same amount that costs increase or determine what portion of the increase they think their customers will pay.

The next-easiest path is to hedge supply price risk through frequent periodic price adjustments. The top three global iron ore producers—BHP Billiton, Vale, and Rio Tinto—replaced previous annual benchmarks with quarterly contract prices in 2011. Vale restructured a quarterly agreement with Japanese Nippon Steel that resulted in a 90% increase in ore contract prices. ArcelorMittal estimated that the increase would raise steel costs by a third in the short term in regions where steel makers have traditionally signed long-term contracts, such as China and Japan.[12]

The third path is to share supply price risk. One way to share risk is to commit to a supplier share of business, which may rise or fall with the underlying volume of business. For example, if there are three target suppliers, a buyer can offer a share to each, subject to each supplier's capacity availability, without committing to the dollar value. Depending on the size of the suppliers, this may still be enough to secure "most-favored nation" pricing (i.e., the lowest that the supplier charges to any of its customers).

The fourth, and usually the most difficult, path is to eliminate supply price and availability risk by either planning far in advance or by integrating vertically with suppliers. Many parts suppliers, which used to plan and schedule on weeks notice, have started planning one year or more in advance. One rotating-equipment parts and repair company plans bulk steel orders for cast parts up to two years ahead and postpones smelting and finishing until it has orders against the material; by machining them to specification when orders come in, the company keeps lead times at two to three weeks, even for precision machined parts. Ordering long in advance and increasing stock levels goes against the grain of lean principles, but many companies have found it necessary to achieve their target service levels.

Owing to strong demand growth, increasing lead times, and scarce raw materials, some suppliers have vertically integrated their supply chains. For example,

- Ruhrpumpen (Mexico) has claimed that its foundry provides a competitive advantage through its guaranteed access to steel.[13]
- BASF built a new plant in Geismar, Louisiana, in 2011, for methylamines that are used in about 20 of the products that BASF manufactures nearby.[14]
- CNOOC bought an 83% controlling stake in Dayukou Chemical in 2007 to extend its supply chain to include direct production of sulfuric acid, which is used as a refinery catalyst or precatalyst and sometimes is used for well acidizing (dissolving dirt and clay to improve flow rates).[15]

To decide which approach is best in each circumstance, some upstream oil and gas companies have instituted market intelligence programs. One company measures the variability of order volume over time. For each major category of purchased materials and services, it assesses the future demand, order lead times, capacity utilization, and prices, quarter by quarter, with forecasts three years forward. It tracks prices and projections to determine the peak of the cycle and proactively work with suppliers to avoid shortages and price spikes.

Managing technology risk

Oil field technology is critical to sustaining production levels while oil and gas reserves are being depleted. The remaining oil is in harder-to-access places, and much of the oil that remains is of lower quality (typically heavier and harder to process). Drilling increasingly involves high-pressure and high-temperature situations. Onshore drilling is increasingly reliant on enhanced recovery, while offshore drilling wears equipment out more quickly owing to the hot and deep (i.e., high-pressure) environment, which also exacerbates corrosion and water-treatment problems. Pumps must increasingly handle more-viscous liquids and slurries. Oil sands and shale bring unique new challenges such as lifting and separation.

Technologies that can overcome these negative cost drivers are therefore extremely valuable to oil and gas producers. One rig rental company signed a deal to place a rig off the coast of Brazil for 950 days at a daily rate of $320,000.[16] Technology saves money and delivers higher levels of output. For example, one major oilfield services firm is using technology to increase yield by leveraging exploration technologies that can help to reach targets faster, using thermal recovery to extract heavy oil, water flooding, and implementing EOR techniques.[17]

Both the costs and the benefits of new technologies evolve rapidly, making long-term technology planning a complex trade-off of uncertain variables. As examples, consider CCS and EOR. Capturing CO_2 emitted from refineries in close proximity to oil production sites and pumping it into nearby oil wells is a relatively new method of EOR and represents an alternative to steam and nitrogen injection. However, the economics of CCS are still being debated. Operations using CO_2 can realize an increase in oil yield of 20% (or even more, according to many estimates), but combining CCS with EOR may be cost-effective only if the infrastructure to transport CO_2 to oil fields is already in place.[18]

For an operator, there are three approaches to managing technology in upstream oil and gas: (1) staying at the cutting edge, (2) delaying implementation of technology until it is mature, and (3) conditionally deciding on a case-by-case or project-by-project basis. In the first of these approaches, technology investments and partnerships are formed to consolidate strengths toward the development and implementation of cutting-edge solutions to upstream technological challenges. For example, Shell and Schlumberger formed a research partnership to increase recovery rates from mature reservoirs at lower cost.[19] The consortium is exploring the use of drill bit sensors to collect more accurate field data, then leveraging that information to customize the bits to meet the requirements of specific fields. Also, Kuwait Oil Company is actively seeking technologies to ensure its future production, particularly in three-dimensional seismic, dynamic pre- and postprocessing, and horizontal/multilateral drilling.[20] Schlumberger established a Research and Geoengineering Center in Rio de Janeiro to more tightly integrate geological science with drilling engineering to improve deepwater drilling yield. Schlumberger expects to employ about 300 scientists and engineers at the complex.[21] General Electric also plans to invest in Brazilian engineering partnerships. It has committed to a $500 million research program, of which $100 million will be dedicated to a new multidisciplinary Global Research and Development Center in Rio de Janeiro for 200 researchers and engineers.[22] The center will focus on electrical generation and control technologies, as well as oil and gas applications.

The second approach, to delay the adoption of unproven technology until it is mature, is the best approach if the technology switching costs and the opportunity cost of lost production are both high. Thus, many oil and gas majors are slow to implement technology, weighing the cost of potential failure against the cost of lost production.

The third strategy, to optimize technology costs and benefits on a project-specific basis (as mentioned earlier, in chap. 1), would be most useful for new industries, such as renewable energy. For companies in these industries, the economics of the key variables are not yet clear—for example, future acquisition cost, the rate of progress of productivity improvements and learning-curve effects over time, operating and maintenance costs, infant mortality rates, and supplier viability risk.

Engineering and Procurement at Minimum Total Cost and Risk

Because of the magnitude of the risk and the interconnectedness of many technological and economic variables, many engineering and procurement contracts are single or dual sourced, often with solution-based or fixed pricing. This favors large service providers.

Single and dual sourcing

Most major oil companies employ a widely used category management approach that consists of category planning, category strategy, negotiation, and supplier performance monitoring. In a series of steps to *define the buy*, they select suppliers, structure agreements, and track and improve supplier performance. As with most strategic sourcing efforts, the goal is to get the most value from suppliers and to rationalize the supply base to strategic partner suppliers.

In upstream, many of the category management analyses often yield the same answer: the economics of complex projects favor one or two suppliers. Single and dual sourcing minimizes the risks of many technology, process, and project management mistakes by ensuring close collaboration and alignment of interests between the owner or operator and the supplier.

Therefore, bid slates often consists of large well-known companies, such as Transocean, Diamond Offshore Drilling, Noble, Ensco, and Seadrill for drilling; Halliburton and Schlumberger for oil field services; and KBR, Saipem, and Technip for EPC services. Even though a large number of possible suppliers are considered, a dearth of technically qualified suppliers often results in a small shortlist of established suppliers.

A tight project schedule increases the tendency to single source. One EPC firm contracted with a single source to build a gas compression and metering station in Asia. Cost imperatives drove the sourcing behavior. The contractor was responsible for managing cost and was held to performance guarantees, all with almost no slack in the budget. Although hundreds of suppliers were interested, the EPC firm quickly selected preferred suppliers and formed a partnership agreement with a single source. The strategy was driven mostly by the tight schedule and the need to deliver long-lead-time compressors. The project was a success and the schedule was maintained because of the close working relationship between the EPC firm, its single source, and the compressor supplier.

Another example of single sourcing was motivated by the technical supply market for seismic surveying, which has few providers. Niche services are a common reason to single source, because suppliers are generally small and highly technical, making economies of scale unlikely and volume leverage impossible. These reasons explain why a major oil company decided to single source deep-sea seismic acquisition and related services.

Another EPC firm, responsible for building an FPSO, dual sourced the job, despite its preference for a single source, owing to the sheer size of the project. Long-lead-time items (huge generators, centrifugal gas compressors, water injection pumps, offloading pumps, and diesel generators) and the *economics of complexity* favored having a single supplier, but the sheer size and capabilities needed to execute the project required more than any one supplier could provide. Consequently, two suppliers were used: One supplier took responsibility for the hull (engineering, procurement, fabrication of modulars, towing, and commissioning); both shared responsibility for the topsides, with one handling the engineering and commissioning and the other handling the fabrication and integration. During construction, the two parties subcontracted to hundreds of other companies and met local-content requirements.

Situations where construction is geographically distributed frequently call for the use of several primary suppliers. For example, one EPC firm chose three to five suppliers in each of four regions to meet its needs for gas/oil separation plants.

Partnering

Because of the economic benefits of maintaining few suppliers mentioned above, both operators and suppliers have become adept at forming partnerships. Partnerships can help achieve tight project schedules by engaging collaborative planning before FID has been reached. One offshore rig contractor set up raw-material purchasing, negotiated an escalation formula for prices and costs at multiple levels of the supply chain, and reserved production slots at a yard in Italy for deployment in Africa—all before the contract was actually awarded. To meet timetables, the group was jointly engineering the project simultaneously with price negotiations.

Recognizing the value of partnerships, some equipment suppliers form dozens of partnership agreements. A compressor supplier uses these to demonstrate its commitment to meeting service and schedule requirements, even when capacity in the industry is tight. Many of its contracts are for 10 years, typically with two renewals of successively shorter terms.

The oil and gas engineering firm Technip entered a long-term strategic partnership with two subsidiaries of Petronas. The firm formalized a collaboration with Petronas subsidiary MISC Berhad—a growing player in the offshore industry, offering floating-facility solutions mainly for FPSOs—to install wellheads and Christmas trees. The partners will work together on onshore and offshore projects, design and build offshore platforms, and exchange expertise. The cooperation with MISC Berhad will enable Technip to deepen its expertise in wellhead and umbilical installation for FPSOs.[23]

FMC Technologies, which makes template manifold systems, wellhead and Christmas-tree assemblies, subsea control systems, and tie-in systems for installation on the seabed, uses partnering to better leverage its total spending and to reduce supplier coordination time. The team's representatives from operations, manufacturing, and engineering share demand forecasts with its suppliers on a quarterly basis, and its innovative "frame agreements" with key suppliers allow it to secure capacity and quality in exchange for a guaranteed share of its purchasing volume. In addition, the company conducts monthly bottleneck analyses.[24] Partly as a result of these efforts, FMC Kongsberg Subsea won Boston Strategies International's annual supply chain award for its superior supplier management, order fulfillment reliability, and overall asset productivity.

Joint process improvement typically yields strong benefits. For example, Amoco's partnership with Red Man Pipe has yielded continuous benefits.[25] Also, Lincoln Electric, a welding-equipment supplier, offers to help its customers identify areas of cost savings through a welding audit. If the client addresses the opportunities, which often involves optimizing electrode and power sources, then Lincoln guarantees a cost savings on its welding products. The joint process helps reduce welding costs.[26]

Specifying and procuring product-service solutions

Increasingly, supply chain transactions consist of solutions, rather than products or services alone. This is part of an overarching trend in manufacturing: "Yesterday you were buying a product for a price. Today you are buying a complete system for a cost," explains an industry association head.[27] Bundling services helps manufacturers to increase exit costs and to maintain or increase their share of the value they generate. Manufacturers that otherwise would have been commoditized by the tsunami of low-cost country sourcing are focusing on a *value-creation triangle* (see fig. 1–10): bundling products and services into solutions, embedding technology, and increasing prices to retain margins. Along with the solution comes a new pricing model—sometimes based on productivity or output (for bits, per foot drilled), rather than a unit price *per each*. For example, drilling is often contracted on a turnkey, footage, incentive, or day-rate basis.[28]

Logistics, including component repair and life-cycle maintenance services, is a ripe area for such solutions. Hence, many 3PLs and 4PLs have emerged to make logistics and order fulfillment more automatic and efficient.

Schlumberger's acquisition of Smith has allowed it to bundle products and services, adding higher value than either products or services would on their own. The joint company can now offer products that support the entire drill string, plus take advantage of Smith's Wilson subsidiary's offering in integrated supply service (i.e., supplier management, inventory holding, field audits, warehousing, and other services such as training, testing, and inspections).[29]

Saudi Aramco bought into the model, selecting Schlumberger to service its new oil field facility in Al-Khafji. Schlumberger opened a $2 million, 215,000-square-foot oil field services facility in Al-Khafji, Saudi Arabia, for repair and maintenance services, and engineering, planning, and operations support. The facility will shorten turnaround times and compete with local third-party maintenance providers.[30]

Scorpion Offshore used NOV's distribution service, called Rigstores, on its rigs, reducing maintenance, repair, and operating costs. Each Rigstore is owned and staffed by NOV and stocked with NOV inventory, allowing Scorpion to focus on its core drilling competencies, instead of on inventory management. It can leverage NOV's supply chain expertise to ensure the necessary products are available on the rig when needed. The services operate on consignment, so the driller pays only for what is used and never deals with obsolete product, slow movers, or write-downs. NOV calculates that the service saves $2.2 million per month for a typical rig.[31]

Other OEM solution offerings have generally met with generally favorable reviews:

- Ingersoll Rand's PackageCare program is a fixed-cost maintenance program that offers customers predictable, scheduled maintenance, and protection from unexpected repair expenses. The company also services third-party equipment under the program.[32]

- KSB provides repair and overhaul services and parts management services (demand planning and warehousing) through Standard Alloys.[33]

- SKF has signed multiple maintenance and inventory management solution contracts.[34]

- Norwegian companies Brønnteknologiutvikling and PI Intervention announced a new common brand, Interwell, to market downhole tools and services under one umbrella. The offering will now provide customers with full life-cycle services from design and prototype through operation and maintenance of downhole tools.[35]

Low-cost country sourcing

Even while upstream oil and gas companies have largely remained wary of low-cost country sourcing for direct purchases, their large suppliers have moved eastward, where capacity and demand are both migrating. Low-cost country sourcing in upstream oil and gas started several tiers back in the supply chain, invisible to the end customers. Top-tier suppliers subcontract to second-tier suppliers, which curtail expense through low-cost country sourcing. For example, ESAB, a welding equipment provider, moved welding equipment production from the United States to Poland to reduce labor costs and tap into growing markets in Asia, Eastern Europe, and the Middle East.[36] PDC inserts (cutters) have moved to China, where one major manufacturer now provides them to nearly all the major bit suppliers, who resell them under their own brand names. The Chinese company operating two 12-hour shifts at two plants to keep up with the demand.

Low-cost country sourcing became more visible when Chinese tubular suppliers gained prominence, although Western manufacturers eventually accused Chinese tubular suppliers of dumping at 30%–90% below cost.[37] Regardless whether this was dumping, the resultant price competition changed the industry structure and pricing behavior, as well as buyers' expectations.[38] Pump manufacturers also went to China in a big way—with both Western manufacturers setting up plants there and local Chinese companies beginning to get supply contracts with those companies. One major pump company's Chinese operation is approved to supply nuclear plants. A top wellhead supplier also went to China, as well as Brazil, Malaysia, Indonesia, and Eastern Europe.

Thus, sometimes unknown to oil and gas customers, suppliers have migrated extensively to China in order to stay competitive. Additional examples of this trend have been seen in cable and membranes:

- In cable, Pirelli expanded into the Asian market by acquiring a majority interest in a joint venture with Nicco; Nexans bought Olex, the largest cable-maker in Australia and the Asia-Pacific region; Alcan is setting up a factory in Tianjin, China; and General Cable acquired Jiangyin Huaming Specialty Cable. In addition, there are more than 1,200 domestic Chinese cable manufacturers.[39]

- In membranes, which are used for water treatment applications in upstream oil and gas, the major suppliers expanded production in Asia to take advantage of lower costs and reduce the cost to serve the Chinese market. GE invested $9 million in its Wuxi rolling facility, and Toray set up a joint venture with BlueStar in Beijing.[40] As with cable and almost every other manufactured product, hundreds or even thousands of emerging local Chinese suppliers will make additional ventures with Western companies and become major independent suppliers.

Rent versus buy decisions

Within each equipment market, unique characteristics affect the decision to buy or rent. For example,

- Most PDC drill bits are now rented, because the bit body can be refurbished by the manufacturer. Operators must return the bit body in good condition, however, or pay a penalty. Roller-cone bits, in contrast, are mostly worn out during the drilling process and therefore are typically not rented.

- The decision to rent or buy downhole tools is driven by the asset's recoverability (i.e., whether it is left in the well), its price, and its level of exclusivity. For example, intelligent controls, while they are expensive and high-tech, are bought instead of rented because they remain in the well. Conversely, shock jars and reamers are usually rented because they are almost always recovered in good enough condition to be reused.

- One drill pipe supplier effectively rents its product. That is, it is the buyer's until they do not want to pay for it anymore; then, the supplier takes it back and rents it out again. Sometimes drilling contractors buy tubulars and rent them to oil and gas operators.

Rentals are taking on a larger share of the total market as buyers try to keep assets off their balance sheets and operators seek to protect new technologies, driven by the need to report strong financial results and minimize their asset bases. For example, IOCs now rent 90%–95% of bits and downhole tools that they need. (In contrast, NOCs often prefer to own the equipment and typically rent only 40%–50%.) The 2008–9 recession accelerated the trend of renting instead of buying, as crude oil prices plummeted from a peak of $147 per barrel (bbl) in 2008 to an average of $43/bbl in the first quarter of 2009. Faced with lower returns on their drilling investments, operators sought to cut costs and reduce risk by renting instead of purchasing. Instead of paying $40,000–$50,000 for a bit, the operator can pay $10,000–$20,000 to rent one for only as long as it needs the equipment.

Suppliers generally prefer to rent because it allows them to protect their intellectual property, compared to selling tools and bits. Another driver for the suppliers is profitability, as the firm gets repeat revenues on the same asset. When the supplier can refurbish a bit body three to four times, it profits more from that asset than it would have if it had sold only one bit.

Construction and Installation

Standardization and modularization

Traditionally, it has been difficult to achieve standardization or economies of scale on very large installations. Rig costs do not exhibit economies of scale, as shown in figure 5–3. In addition, complex machinery and equipment, such as compressors and turbines, requires customization to each application to function at peak performance in each operating environment. Given the operating cost and the opportunity cost of lost production, such customization is usually considered a small price to pay for high operating performance and production rates.

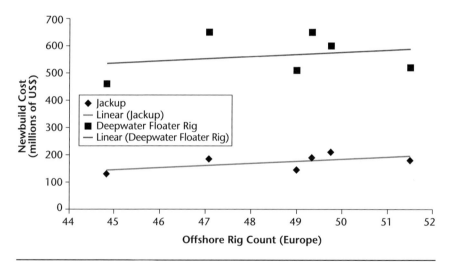

Fig. 5–3. Jackup and deepwater floater rig economies of scale (*Source:* author's analysis of data from *Rigzone*, February 15, 2010.)

Nonetheless, oil and gas companies and their contractors have achieved major cost reductions through standardization of processes and modularization of large installations during the engineering, construction and installation phases. Statoil, FMC Kongsberg Subsea, and Aker Solutions cooperate in standardizing specifications for subsea workover systems. Multifunctionality for workover systems means the same system can be used independently of supplier of subsea infrastructure. This results in cost savings and simplified well maintenance. In addition, one supplier can work on another supplier's equipment, which provides a greater degree of safety.

BP used standardized processes to build colossal offshore platforms in the Azeri-Chirag-Gunashli field in the Caspian Sea. It set up a production line and standardized offshore jackets and floatover topsides. By building onshore and transporting to its offshore resting place, BP was able to take advantage of both lower construction costs and economies of scale through standardization. It plans to use the concept as a standard practice in the future.[41] The topsides—wellbay, process, utilities, drilling, and quarters— were identical and modular. Each jacket substructure was built on the same concept. The first phase was six weeks ahead of schedule, and the second was 4–5 months early. The design person-hours per ton of topsides decreased from 42 in the first phase (Central Azeri) to 31 in the second phase (East Azeri) to 11 in the third phase (East Azeri). The design person-hours per ton for drilling decreased from 50 to 10 between phase 1 and phase 2. The cycle time from cutting first steel to a completed deck went from 33 months in phase 1 to 31 months in phase 2 and is targeted at 29 months for phase 3. Person-hours required for completing and testing dropped from 900,000 to 800,000 early in phase 2 and, subsequently, to 500,000 in phase 2. Jacket building time dropped from 18 months to 9 months, and floatover deck installation fell from 56 hours to 35 hours, while jacket installation time fell from 37 days to 29 days.[42]

Shell was able to reduce drilling days from 60 to 25 and to reduce drilling costs by 30%, on the Pinedale shale gas field in Wyoming, which involved hundreds of multistage, fractured wells. It reduced completion costs by 60% and cycle time for completion from one well every 60 days to five wells in 20 days. On the Groundbirch shale gas field in British Columbia, its first well took 40 days, compared to 15 days three years later. In the Groningen gas field in the Netherlands, where Shell built seven compression stations to maintain adequate pressure for drilling, it was able to use the same construction team, design, and methodology to achieve 20% lower costs and cut 10%–15% from the project cycle time by the third station. The company estimates savings up to 35%–40% in project cost and 15%–40% reduction in project cycle time, through standardization.[43]

Should-cost analysis

Due in part to the small number of qualified suppliers and the industry concentration previously referenced, market prices sometimes diverge from competitive levels over time, and structured initiatives to realign them such as strategic sourcing programs and e-sourcing platforms can yield big benefits for buyers. For example, Petronect, the e-procurement portal for Petrobras, implemented the SAP Supplier Relationship Management application and the SAP NetWeaver Portal component, which cut the average closing price for bids by 22% and reduced operational costs as a percent of sales by 10%.[44]

In a more targeted way for specific bids, should-cost analysis can provide a useful reference point for bids when only one or two suppliers qualify. The analysis steps are as follows:

- Tailor to specific orders
- Start with a base period price by using cost inflation factors
- Update to current prices
- Subtract excess profit
- Subtract cost of overengineering
- Subtract diseconomies of scale
- Subtract diseconomies of scope
- Subtract cost of coordination
- Commit to a long-term agreement

In the following example, the process starts with a 2008 price and cost breakdown for the exact specification of the desired equipment, including profit margin (table 5–1). That price is then inflated to the current price by using an inflation factor for the subcategory, such as centrifugal engineered oil field compressors. After inflation for the category, the cost factors for the unique specifications of the equipment can be modeled. The inflation factors take into account the capacity utilization of the supplier and any premium that may be attached owing to excess demand, which is measured by current versus historical lead time reference points.

Table 5–1. Should-cost breakdown of cost structure (illustrative and partial)

Index Name	Cost Grouping
Profit margin	Profit
Electrical equipment manufacturing	Labor
Professional wages and salary	Labor
Capital equipment	Materials
Benefits	Labor
Component manufacturing	Labor
Forged parts	Materials
Anti-corrosion alloys	Materials

The resulting analysis produces a time series of capacity utilization, lead time, and prices for the specific equipment (fig. 5–4). By application of algorithms based on the differences in the shape of the capacity utilization, lead time, and price curves, it is possible to create a profit margin adjustment factor. If the market is in an *up* mode, then the profit margin adjustment factor inflates the prices; if the market is in a *down* mode, the dominant price index adjusts downward (and possibly deflates) the historical price. In the model, prices are upwardly flexible and downwardly sticky (fig. 5–5).

Price Index

2007 Q2	2007 Q3	2007 Q4	2008 Q1	2008 Q2	2008 Q3	2008 Q4	2009 Q1	2009 Q2	2009 Q3	2009 Q4	2010 Q1
132.91	132.98	133.89	134.68	135.31	135.65	136.36	136.95	137.66	138.07	138.36	139.19
118.30	118.71	119.14	120.12	120.42	120.74	121.37	121.67	122.46	123.20	123.21	124.09
123.25	124.21	124.63	125.49	125.52	125.65	126.18	126.39	126.52	127.35	127.47	127.70
122.40	122.55	122.70	123.53	123.91	124.74	125.53	125.88	126.85	126.94	127.12	127.75
170.60	171.06	171.99	172.47	172.89	173.10	173.43	174.18	175.04	175.55	176.51	176.76
132.59	133.36	134.25	134.95	135.13	135.71	135.86	136.38	137.35	137.71	137.88	138.12
126.69	127.20	128.16	128.57	129.08	129.44	129.69	130.23	130.42	130.62	130.93	131.68
104.05	104.98	105.84	106.48	106.87	107.84	107.92	108.43	109.26	110.23	111.21	111.72
127.14	127.20	127.79	127.87	128.55	129.07	130.07	130.12	130.94	131.82	131.84	132.38
108.80	109.48	110.31	110.38	111.22	111.28	111.42	111.93	112.72	113.69	114.00	114.30
114.34	114.64	115.46	115.74	116.47	117.12	117.43	117.81	118.34	118.82	118.86	118.93
114.09	115.05	115.36	115.75	115.97	116.76	117.07	118.01	118.79	118.87	119.13	119.77
188.06	189.00	189.26	189.69	190.25	190.30	190.59	190.99	191.30	192.24	193.13	193.79
128.90	129.44	129.47	130.28	131.02	131.86	132.42	133.03	133.12	133.52	133.77	134.47
116.19	116.34	116.77	117.05	117.57	118.39	118.42	119.03	119.40	119.95	120.36	120.97

Fig. 5–4. Should-cost updating of base year cost (illustrative)

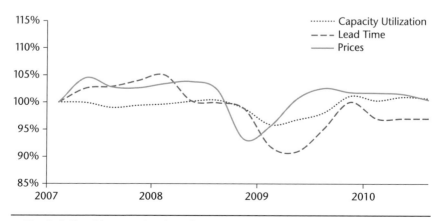

Determination of Profit Margin Adjustment Factor Based on Cost and Price Trigger Factors

2007 Q2	2007 Q3	2007 Q4	2008 Q1	2008 Q2	2008 Q3	2008 Q4	2009 Q1	2009 Q2	2009 Q3	2009 Q4	2010 Q1
2.92%	-0.23%	-1.41%	1.06%	3.48%	1.37%	-4.53%	-4.16%	-0.21%	2.57%	2.06%	1.44%
1.16%	-2.2							2.84%	-0.55%	1.89%	4.20%	4.38%
3.64%	5.6											

Price Index Forecast Computation

2007 Q2	2007 Q3	2007 Q4	2008 Q1	2008 Q2	2008 Q3	2008 Q4	2009 Q1	2009 Q2	2009 Q3	2009 Q4	2010 Q1
1.47%	1.15%	1.50%	2.34%	2.16%	1.81%	-0.93%	-1.16%	-0.34%	1.51%	1.58%	1.60%
1.79%	0.73%	0.92%	2.82%	3.29%	3.11%	-1.96%	-2.37%	0.53%	1.11%	1.90%	1.48%
3.63%	3.03%	2.81%	3.01%	2.84%	2.26%	0.11%	-2.98%	-1.47%	1.30%	0.44%	4.29%
Yes	Yes	Yes	Yes	Yes	Yes	Yes	Yes	No	Yes	Yes	Yes
1.47%	1.15%	1.50%	2.82%	2.16%	3.11%	-1.96%	-2.37%	-0.34%	1.51%	1.90%	1.60%
Cost	Cost	Cost	PPI	Cost	PPI	PPI	PPI	Cost	Cost	PPI	Cost

2007 Q2	2007 Q3	2007 Q4	2008 Q1	2008 Q2	2008 Q3	2008 Q4	2009 Q1	2009 Q2	2009 Q3	2009 Q4	2010 Q1
0.87%	1.29%	2.35%	3.47%	1.79%	2.26%	-0.14%	-0.05%	-0.86%	1.55%	2.22%	2.48%
1.16%	-2.23%	0.61%	4.02%	4.37%	7.41%	0.46%	2.84%	-0.55%	1.89%	4.20%	4.38%
3.04%	4.02%	2.28%	2.82%	1.21%	3.20%	0.75%	-5.20%	-3.71%	-2.24%	-1.10%	2.81%
Yes	No	Yes	Yes	Yes	Yes	No	No	Yes	Yes	Yes	Yes
Yes	No	Yes	Yes	Yes	Yes	No	No	Yes	Yes	Yes	Yes
0.87%	1.29%	2.35%	4.02%	1.79%	2.26%	0.46%	-0.05%	-0.55%	1.55%	4.20%	4.38%
Cost	Cost	Cost	PPI	Cost	PPI	Cost	PPI	PPI	Cost	Cost	PPI

Fig. 5–5. Should-cost computation of excess profit margin (illustrative)

Leads, lags, and differences in the shape of the lead time, capacity utilization, and price curves shed valuable insight on the future direction and magnitude of price changes. For example, in figure 5–6, in the first and second quarters of 2009 suppliers raised prices before capacity utilization had returned to normal levels, indicating a period of "softness," during which sellers could successfully make additional profits and buyers could successfully negotiate prices down.

Fig. 5–6. Should-cost estimation of current market price premium, 2007–10 (illustrative)

The should-cost methodology should also take into account the following:

- **Opportunities for standardization or product simplification.** Often, highly engineered rotating equipment (e.g., compressors and turbines) must be customized for each application to remain cost-effective over their life spans. However, some equipment (e.g., certain types of pumps) may be able to be standardized, as to an ANSI specification, at significant cost savings.

- **Economies of scale—from volume purchasing.** Unit costs and prices can decrease as the number of units increases. For example, one company's power units became less costly the more the company produced. In 2008 the company produced units with a total capacity of 22 megawatts (MW), at an average cost of $3 million/MW, whereas in 2010, the company produced 52.8 MW in total of units at an average cost of $2.4 million/MW—an 18% reduction in cost per MW. Also, sometimes efficiency improves when joining multiple smaller units because of the ability to flex capacity on and off as needed.

- **Economies of scope—from purchasing multiple types of products or services from the same supplier.** For example, Several of the major electrical equipment manufacturers supply motors, electrical distribution and control equipment, and lighting, but few buyers think to leverage their purchasing power by consolidating their purchases of transformers and lighting. Sometimes suppliers consolidate their services before buyers do. For example, fire suppression and security detection companies, such as Tyco, have recently combined their offering under a *life-safety* umbrella, for both cost and marketing benefits.

- **Potential savings from more coordinated production and inventory planning.** By collaboration with suppliers on production planning, inventory stocking and warehousing, and logistics, further savings may be possible.

The net result of the layered analysis is a should-cost price (fig. 5–7).

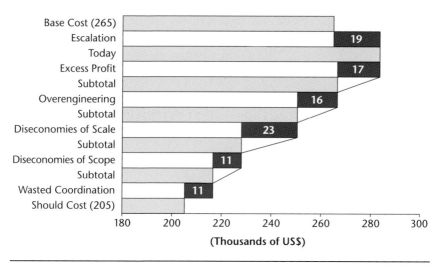

Fig. 5–7. Should-cost waterfall chart (illustrative)

Further avenues to cost reduction in the engineering, construction, and installation phases include substitute materials, global sourcing, and purchasing directly from suppliers (*tier-skipping*), discussed in the next sections.

Substitute materials

Substitution of consumable materials, even a slight change in the specifications, at the design stage can save significant cost, since much of the ongoing operating and maintenance cost is embedded at the design stage, and hard or uneconomical to change thereafter. One example of a material substitution trade-off is in drilling fluids. There are many choices, including water-based and oil-based muds, as well as in terms of specific ingredients such as barite and bentonite, to achieve the required characteristics and wellbore reactions. Within this framework, the grade of mineral may offer cost versus volume trade-offs. One conservation method employed to extend the supply of barite has been to lower the required grade of barite in the fluid mix from 4.2 to 4.1 specific gravity. Essentially this lowers the ratio of barite to water, stretching the supply of barite with little to no adverse affects for drilling under some circumstances.[45]

Alternatives to barite in the drilling mud include hematite, celestite, ilmenite, and iron ore. However, as they are more abrasive and more expensive, these substitutes have not been widely adopted. Hematite, the primary alternative to barite, is more abrasive, so it wears out equipment (e.g., pumps) faster. It also has a higher specific gravity (>5), which would require operators that use it as a substitute to revise mud properties and system designs to compensate.[46]

Tier-skipping

At the procurement and construction phase, operators must evaluate the pros and cons of opportunities to source directly from subsuppliers, a practice known as *tier-skipping*. For example, some operators are now buying fracturing sands (i.e., proppant) from a materials science company instead of through distributors. The company's technologically engineered proppant lasts longer and requires less additive compared with sand for shale (or tight sands) wells. Given rapidly rising prices and the perceived lack of value addition by distributors, the company and its customers are giving the middleman the boot.

Also, a minerals supplier that once used to sell only to drilling-mud companies is now offering to sell directly to operators who want to mix their own drilling and completion fluids. The company has been able to achieve economies of scale because it supplies minerals to many different industries. Those economies of scale allow it to offer advantageous terms to operators.

Operations and Maintenance Cost Reduction

In some ways, operations and maintenance in upstream oil and gas operations is just like in other industries. However, it differs in its unusually strict safety standards, demand volatility, intermittent and high-value inventory demands, the very large size of most project cargo, the specialized nature of blending and transporting sometimes hazardous liquid bulk, and the challenging quality standards. Accordingly, this section will focus on these points of differentiation, covering

- Health and safety priorities as they extend throughout the supply chain

- Production scheduling optimization, including flexible manufacturing

- Logistics and inventory management for stochastic demand

- Asset tracking

- Transportation optimization, including stochastic inventory management, outsourcing, and managing 3PLs and 4PLs

- Quality standards and supplier quality management

Health and safety priorities throughout the supply chain

On April 20, 2010, the explosion at BP's Macondo well not only cost 11 lives but may result in total monetary costs of as much as $60 billion (including clean-up and maximum penalties). The lack of risk management practices and designated accountability—and inadequacy of risk management alignment among BP, Cameron, Halliburton, Transocean, and the U.S. Minerals Management Service [MMS]—is instructive. ISO 31000:2009 requires that risk management be integrated into all organizational procedures for particular types of risk or circumstances. According to these guidelines, each of the five companies needed to have in place the following:

- A proper risk management policy

- Authorized representatives on the project who were responsible for effectively implementing and analyzing the risk involved at various stages

- Procedures, guidelines, and standard practices for managing such a crisis

Risk management policy. MMS was responsible for establishing directions and a deterrent approach to risk management for organizations working on offshore facilities. It gave waivers for legally required environmental impact tests, and investigators later found that it did not have sufficient staff or auditors to monitor the risk of the project, which could equally have led to crisis on any other facility by other operators, such as Shell and ExxonMobil, among others.

Risk analysis. BP skipped a time-consuming cement log test that would have determined if there were any weaknesses in the cement. Cameron manufactured the blowout preventer (BOP) that did not activate, while Halliburton laid the cement plugs in the well that yielded under pressure. The spill might have been suppressed earlier if the authorities comprehended the level of risk and depth of the disaster by assessing all the risk parameters throughout the drilling life cycle.

To evaluate the probability of the Macondo spill and its magnitude, during the planning phase, the MMS utilized the *oil-spill risk model* developed by the U.S. Geological Survey. This model relied on previous incidents, using historical data as the basis for a low probability of a spill. However, it did not detect and isolate specific faults at an early stage, before they caused a problem, so assigned only a 5% probability to large oil spills. The model estimated the most likely size of a large spill at 4,600 bbl, as compared to the actual figure of 4,900,000 bbl.[47] During the three-month period of the spill, the magnitude of the disaster changed significantly with the estimated flow rate increasing from the initial 1,000 bbl per day (bpd) to a final official figure of 62,000 bpd. Hence, during the Macondo oil spill, as assessments by regulatory bodies varied greatly over a period of time, the risk evaluation process turned questionable.

Standard risk management practices and procedures. Analysis of the events suggests that the Macondo crisis could have been avoided had all five parties embraced well-defined frameworks, processes, and practices to control risk. The documented guidelines for continual checking, monitoring, critically analyzing, or specifying the status throughout the drilling life cycle could have been effective.

Neither BP nor Transocean established a sufficiently formal risk treatment plan to deal with a crisis of this proportion. ISO 31000 standards define risk-mitigation guidelines to remove the risk source and establish methodologies to mitigate the consequences, but in the case of the *Deepwater Horizon* oil spill, the risk-mitigation strategy was adopted *after* the incident, resulting in a fragmented approach that relied on hasty decisions. In the end, the skimming and booming methods used on the spill only recovered about 10% of the total oil that was spilled.[48] In fact, the use of dispersants at the spill site impeded additional recovery. Not only did BP not have proven chemical technology to disperse oil at such a large scale or at water depths of thousands of feet but the actual measures to be taken in case of an oil spill were never clearly outlined.

Since the Macondo crisis in 2010, well operators and lessees are being held to higher standards. Well operators now need to submit detailed emergency action plans on a well-by-well basis, and stock the required safety equipment on the rig for immediate use in case it is needed. For example, in the United States, the Bureau of Ocean Energy Management, Regulation and Enforcement (BOEMRE) has introduced the following rules for improving safety and accident prevention, blowout containment, and spill response:

- A *Drilling Safety Rule* imposes higher standards for well design, casing and cementing, and well control procedures. Operators must get independent third-party inspection and certification of the proposed drilling process. In addition, an engineer must certify that the BOPs are capable of severing the drill pipe in the environment and service where it will be placed.

- A *Workplace Safety Rule* requires operators to identify risks, establish barriers to those risks, and actively and methodically reduce human and organizational errors. Operators now are required to develop a life-cycle safety and environmental management program that identifies the potential hazards and risk-reduction strategies from drilling through decommissioning.

- A *new directive (BOEMRE NTL-N06)* requires that operators develop oil spill response plans, including a well-specific blowout and worst-case discharge scenario, and support their figures with the underlying assumptions and calculations.

After the *Deepwater Horizon* disaster, BP also self-imposed standards for its equipment, especially BOPs. The new standards include enhanced response measures for blowouts and oil spills and require that an independent third party test the cement used for deepwater well sealing. They call for a BOP with at least one set of *blind rams*, which will prevent a section of drill pipe from blocking open the wellbore. It requires that a third party perform the testing and maintenance on a subsea BOP whenever it is brought to the surface. BP also included intensive measures in its Oil Spill Response Plan, a set of rules related to well design, spill containment, and modification in BOPs, based on lessons learned from the Macondo disaster.

Because of the evident possibilities for disastrous accidents in upstream oil and gas, quality standards are critical throughout the entire oil and gas value chain. Equipment suppliers need to follow rigorous quality regimes as much as do operators. For example, compressor manufacturer Ebara implemented failure mode/effects analysis in new product design, the individual production stages, and developed a work standard to prevent nonconformities, including the five "why's" of lean methodology. Ebara is also continuing the lean journey toward lower inventory and lower cost through its Mindora Campaign and M Zero Challenge, both of which foster collaboration between the design, manufacturing, and sales departments.[49]

Lean and flexible production management

Since upstream operations are asset intensive—rigs can cost several hundreds of thousands of dollars per day, and subsea and deepwater operations further exacerbate asset intensity—shortfalls in productivity can rapidly erode the financial return on those assets. Worn drill bits reduce productivity, to the point where the cost in rig time rapidly exceeds the cost to buy a new bit.

Production scheduling optimization can increase the efficiency of upstream assets, simultaneously helping to reduce costs and increase throughput (revenue and profit). Thailand's PTT Group, which has operations in upstream oil and gas, deployed a software application to simulate Unit 5 of its gas separation plant so that it could increase production capacity. AspenONE Engineering and

Aspen HYSYS created a process simulation model of the ethane recovery unit. The initiative optimized the unit, increasing plant capacity to 106% of design capacity. This led to an improvement in daily profitability of $46,000/day.[50]

Flexible manufacturing and assembly is a powerful tool for dealing with demand volatility. As noted in the constraints management section of Part 1, reducing cycle time in every process along the value chain makes the entire system more able to adapt to changes in demand. One way to operate lean is to make rapid changeovers—for example, quickly deploying drilling rigs or installation vessels to where they are needed. Statoil reduced lead time for subsea tie-backs by 50%, decreased the capex by 30%, and reduced the time from discovery to production from five years to 2.5 years.[51] The company is using standardization as a way to make smaller reserves profitable. Exxon Angola's Kizomba floating production project experienced a five-month shorter cycle time for the second than for the first project: Kizomba A took 36 months, whereas Kizomba B came onstream after 31 months of construction.

At the component level, pump suppliers are implementing mechanisms to meet changing demand more economically and responsively. Flowserve added second and third shifts during the economic turmoil that followed the financial crisis to increase its ability to flex capacity quickly without adding real estate or selling plant assets,[52] an extension of its normal procedure of shifting production globally to balance its capacity utilization across plants. ITT's modular assembly operations help it scale production capacity to actual demand, as well as to reduce its order cycle times. In addition, ITT builds its modules on skids that can be easily transferred across product lines; this enables it to move the product to whichever production operation has the lowest utilization rate, thereby balancing capacity. Balanced capacity allows it to lower the average capacity level of all of its operations since there are fewer and smaller peaks.[53]

Logistics and inventory management for stochastic demand. How do you formulate a stocking policy for a product that is needed in enormous quantities only at unpredictable times? During the Macondo oil spill, BP ruled out the majority of dispersant suppliers simply because of a lack of capacity to produce the volumes it needed. The dispersant market is small, and Nalco was chosen as the sole supplier because of its ability to quickly reallocate resources to produce mass quantities of the proprietary solvent Corexit.[54]

Specialized inventory management solutions exist for intermittent demand. The underlying replenishment statistics rely more on the probability of equipment failure, as is common in capital spares, than on replenishment, as is common in typical fast-moving goods applications. Analyses based on mortality rates are used by analyzing Weibull distributions rather than analyses based on z statistics, which assume a distribution around a normal curve.

Other, process-based approaches can also deal with unpredictable demand. For example, lean manufacturing, lean distribution, and other derivatives of lean thinking reduce buffers. Consequently, this reduces the chance of getting stuck with a lot of inventory that is not needed or, conversely, of coming up short when demand spikes.

Refurbishment has become a popular alternative to replacing aging equipment because of its shorter lead times. For this reason, when supply was tight in recent years, many operators turned to refurbished or used equipment rather than ordering new equipment. For example,

- Hitachi Cable developed a continuous processing technology for reusing silane cross-linked polyethylene waste as insulating material in new power cables.[55]

- NSK reconditioned and refurbished bearings rather than replacing them, to save a paper mill $75,000 in capex.[56]

Transportation of oversized project cargo. The oversized items result in unusual *project cargo* requirements that few transport providers can handle. Usually, operators contract with specialized transporters with the right equipment for the job. For example, for its Azeri, Chirag, and Gunashli development in Azerbaijan in the mid-2000s, BP contracted Bertling Global Logistics to transport over one million tons of equipment and materials, making about 23,000 individual freight movements.[57]

Petrobras converted a tanker into an FPSO for operation in the Cascade and Chinook fields in the Gulf of Mexico. Altus Oil & Gas loaded and shipped the FPSO topsides, as well as turret and mooring system, for deliveries to Keppel Shipyard in Singapore and onward to an offshore site in the Gulf of Mexico. The submerged-turret production buoy and turret weighed 1,200 tons and was about 30 meters tall, so lifting the buoy and finding the right type of vessel to move it involved specialized engineering and planning, especially to maintain a safe and stable center of gravity during ocean transit. The team's analysis led to the decision to use an adapted, less expensive heavy-lift vessel rather than a specialized vessel.[58]

Logistics outsourcing to 3PLs and 4PLs. Many operators have concluded that the nature of logistics competency required in order to satisfy such unpredictable demand and complex shipments is not a core competency, and have consequently turned to integrated logistics providers to ensure that their target levels of service are met. The arrangements have covered a spectrum from project shipping contracts to full-scale outsourcing of transportation and logistics planning and management.

On the more transactional level, operators hire specialized haulers for single shipments and use conventional freight forwarding services. For example,

- Using a conventional freight forwarding model, Statoil (Orient) used Shenship Logistics through the 1990s and 2000s to handle shipping and freight forwarding. Among its services, Shenship shipped Christmas trees door-to-door from a supplier in Singapore to Statoil's facility in Chiwan, China.

- Similarly, KNPC awarded Agility a contract for door-to-door freight services, customs clearing, and transportation for five years.

On the more strategic level, operators hire 3PLs with intentions of achieving certain goals, such as specific cost-savings targets:

- Saudi Aramco established its Electrical Stocker Distribution Program, an outsourced model that represents a new business model in the Middle East, to rationalize cost and embed innovation in the electrical equipment supply chain. The $275 million program is designed to simplify ordering processes, lower costs, and improve fill rates.[59]

- Petrom, the Romanian oil company, chose Tenaris to manage its inventory in the Black Sea, to reduce carrying costs and equipment damage. Petrom, the largest oil and gas producer in Southeastern Europe, currently has one year's worth of tubulars on hand as a result of the agreement but does not have to pay to maintain it.[60]

- Traverse Drilling outsourced its sourcing, purchasing, industrial supply, logistics, and supply chain management activities with the goal of increasing its operational efficiency by streamlining supply and service and allowing it to focus on its core competency (i.e., drilling).[61]

- A major oil field equipment supplier and a major compressor component manufacturer both hired CEVA Logistics to manage their spare-parts distribution operations.[62] The oil field equipment supplier estimated savings of $50 million from the outsourcing deal.

- Nations Energy chose the Wood Group to provide supply chain management services. Under a $10 million, two-year contract, procurement, inventory, and logistics management services in Aktau, Kazakhstan, are covered.[63]

Asset tracking. An asset-tracking initiative or technology investment can pay off if the cost of error and subsequent redeployment (manufacturing, transportation, warehousing, handling, etc.) is high.

The tagging of drill pipe at Petrobras is a good example. To ensure that drill pipe is in the right place at the right time and in the right quantity, Petrobras required Weatherford to track all the items that it ships to its offshore platforms. Weatherford tagged 7,500 segments of drill pipe, with the help of Trac ID Systems, a Norwegian RFID tag provider. Workers tag pipe sections in the field and key in data such as the location and whether there were any defects or damage to the pipe. With the information gathered via the RFID solution, Weatherford can set proper stocking levels, track inspection histories, and monitor condition in real time. In addition, Weatherford can *peg* (identify and trace to a certain job or lot number) the location of pipe while it is en route from its stocking point to the offshore rigs. For this application, Weatherford is using ISO 18000-2 compliant one-inch wide 134.5-kilohertz tags.[64]

Asset tagging can also pay off by communicating with remote devices to increase production and yield. Weatherford and Marathon Oil are using RFID to remotely open and close the cutter blocks on reamers, thereby increasing well productivity. The tag is positioned in the middle of the inside of the drill stem, while the reader is positioned on the drill stem itself. The result is more reliable than hydraulically controlled reamers, which have a higher risk of damage and failure because of their manual connection from the wellhead to the reamer. Weatherford named the system RipTide. Marathon Oil, which bought RFID technology provider In-Depth Systems in 2001, also uses RFID in coiled tubing, packer setting, wellbore cleanup tools, zonal isolation, perforation, and cementing. M-I SWACO also uses RFID for drilling and wellbore cleanup applications,[65] and GE and Baker Hughes are offering ancillary monitoring and control products that allow wireless remote monitoring and Web portal access to product inventory status.[66]

Asset tagging can also improve compliance and safety. Halo, a manufacturer of ropes, chains, and similar fastening equipment, is using RFID to warn of impending safety inspection requirements on safety-critical slings.[67]

Notes

1 Raymond, Martin S., and William L. Leffler. 2005. *Oil & Gas Production in Nontechnical Language*. Tulsa, OK: PennWell.

2 Graham, Mark, Mark Cook, and Frank Jahn. 2001. Project and contract management. *Hydrocarbon Exploration and Production*. 55.

3 Raymond and Leffler, 2005.

4 Seddon, Duncan. 2006. *Gas Usage and Value: The Technology and Economics of Natural Gas Use in the Process Industries*. Tulsa, OK: PennWell. P. 12.

5 Wright, Charlotte J., and Rebecca A. Gallun. 2008. *Fundamentals of Oil & Gas Accounting*. 5th ed., Tulsa, OK: PennWell.

6 Jacoby, David. 2010. Balancing economic risks: Tips for a well-structured deal. *Middle East Energy*. September. P. 5.

7 Ebrahimi, S. N. 2010. Maximising the Value of Strategic Partnerships: NOCs & IOCs. Paper presented to the National Oil Companies Congress, London. June 21.

8 De St. Aubin, Mark C. 2003. Rebuilding Iraq: How much risk is too much? *Trust the Leaders*. Fall (5). http://www.sgrlaw.com/resources/trust_the_leaders/leaders_issues/ttl5/917/ (accessed May 19, 2012).

9 Copeland, Thomas E. 2001. *Real Options: A Practitioner's Guide*. Texere. New York: LLC.

10 Boston Strategies International.

11 Topf, Andrew. 2011. Vale bends to new quarterly iron ore pricing model. Mining.com Web site. http://www.mining.com/2011/12/08/vale-bends-to-new-quarterly-iron-ore-pricing-model (accessed January 30, 2012).

12 Ibid., p. 1.

13 Ruhrpumpen. [n.d.] Ruhrpumpen Foundry Fundemex. Ruhrpumpen Web site. http://www.ruhrpumpen.com/files/Brochures_Generales/Fundemex.pdf (accessed May 19, 2012).

14 Basf. 2009. BASF to build new methylamines plant in Geismar, Louisiana. Press release, May 18. http://www.basf.com/group/pressrelease/P-09-241 (accessed May 19, 2012).

15 *Sulfuric Acid Today*. 2007. China's CNOOC to buy Cinda AMC's Hubei Dayukou Chemical stake for 530 mln yuan. *Sulfuric Acid Today*. December 17. http://www.h2so4today.com/publish/posts/1/chinas-cnooc-to-buy-cinda-amcs-hubei-dayukou-chemical-stake-for-530-mln-yuan.html (accessed May 19, 2012).

16 German, Erik. 2010. Rent this oil rig for just 30 million bucks! *Global Post*. February 3. http://www.globalpost.com/dispatches/bric-yard/rent-oil-rig-just-30-million-bucks (accessed February 11, 2011).

17 Duroc Danner, Bernard J. 2010. The role of oil field services in the energy dynamic. Paper presented at the National Oil Congress, London, June 22.

18 U.S. Department of Energy. 2009. Secretary Chu announces first awards from $1.4 billion for industrial carbon capture and storage projects. Press release, October 2. http://energy.gov/articles/secretary-chu-announces-first-awards-14-billion-industrial-carbon-capture-and-storage (accessed May 17, 2012).

19 Tendersinfo News. 2010. Shell, Schlumberger in joint oil & gas multi-year research technology cooperation. *Tendersinfo News*. December 20.

20 KOC. 2010. Strategies to sustainably fulfill energy needs. Paper presented at the National Oil Congress, London, June 22.

21 Schlumberger. 2010. Schlumberger inaugurates Brazil research and geoengineering center. Press release, November 16. http://www.slb.com/news/press_releases/2010/2010_1116_brgz_slb.aspx (accessed May 19, 2012).

22 General Electric. 2010. GE to invest $500 million in Brazil for accelerated growth. Press release, November 10. http://www.google.com/url?sa=t&rct=j&q=&esrc=s&source=web&cd=2&ved=0CDQQFjAB&url=http%3A%2F%2Fwww.genewscenter.com%2FPress-Releases%2FGE-TO-INVEST-500-MILLION-IN-BRAZIL-FOR-ACCELERATED-GROWTH-2cab.aspx&ei=gC83T7vvGKPJ0AGOo6jPAg&usg=AFQjCNEvOkZGQWq3BoE2Jij15wdRhnqltA&sig2=InYId6__dTPlFbZnu-LEAQ.

23 Technip. 2010. Technip and PETRONAS subsidiaries agree to establish a strategic business collaboration. Press release, August 20. http://www.technip.com/en/press/technip-and-petronas-subsidiaries-agree-establish-strategic-business-collaboration.

24 FMC Technologies. 2010. FMC Technologies: Transforming planning and scheduling processes. Supply Chain Council Web site. http://supply-chain.org/civicrm/event/info?id=183&reset=1 (accessed May 19, 2012).

25 Underhill, Tim. 1996. *Strategic Alliances: Managing the Supply Chain.* Tulsa, OK: PennWell. 1996.

26 Lincoln Electric. [n.d.] Guaranteed cost reduction. Lincoln Electric Web site. http://www.lincolnelectric.com/en-us/company/custom-solutions/Pages/guaranteed-cost-reduction.aspx (accessed May 19, 2012).

27 Gracia, Bernard. 2010. How to measure procurement performance: an holistic approach. Presentation to the 5th Annual Global Procurement and Supply Chain Management for the Oil and Gas Industry, Barcelona. September 27. For additional insights on solution buying, see Jacoby, David, and Bruna Figueiredo. 2008. The art of high-cost sourcing. *Supply Chain Management Review.* May/June.

28 Graham, Cook, and Jahn, 2001, p. 61.

29 Shannon, Kerri. 2010. Schlumberger's acquisition of Smith the latest evidence of a takeover trend. *Money Morning.* February 22. http://moneymorning.com/2010/02/22/schlumberger-smith-takeover/ (accessed May 19, 2012).

30 Al Mashni, Rima Ali. 2010. Schlumberger opens Al-Khafji oilfield services base in Saudi Arabia. AME Info. June 22. http://www.ameinfo.com/235993.html.

31 Mathilakath, Santosh, and Jonathan Rhoads. 2008. Rigstore supply model aims to cut MRO costs. *Offshore Magazine.* April 1. http://www.offshore-mag.com/articles/print/volume-68/issue-4/drilling-technology-report/opened-for-business._printArticle.html.

32 Ingersoll Rand. 2008. Ingersoll Rand signs 4,000th unit under its PackageCare™ service agreement. Press release, October 13. http://www.ingersollrandproducts.com/news/article.aspx?news_id=155 (accessed May 19, 2012).

33 *WaterWorld.* 2010. KSB buys US pump service company. *WaterWorld.* August 5. http://www.waterworld.com/index/display/article-display/6656514988/articles/waterworld/drinking-water/distribution/2010/08/KSB-buys-US-pump-service-company.html (accessed May 19, 2012).

34 SKF. 2010 Annual Report. Pp. 15–16.

35 Interwell. 2011. WTG, BTU and PI Intervention will now approach the market as one voice, with a new name and identity: Interwell. Press release, March 30. http://www.interwell.com/news/becoming-interwell-article81-120.html (accessed May 19, 2012).

36 ESAB. [n.d.] ESAB Cutting Systems: 1996 – 2008. ESAB Web site. http://www.esab-cutting.com/index.php?id=707 (accessed May 19, 2012).

37 Targeted New Service. 2011. Commerce finds dumping and subsidization of drill pipe from the People's Republic of China. Targeted News Service, January 4.

38 Roknick, Michael. 2009. U.S. pipe makers prevail. *Sharon Herald*. September 9. http://sharonherald.com/local/x1081150087/U-S-pipe-makers-prevail (accessed May 19, 2012).

39 PR Newswire. 2007. Alcan breaks ground for its new world-class facility in China. The Free Library, August 7. http://www.thefreelibrary.com/Alcan+ breaks+ground+for+its+new+world-class+facility+in+China.-a0167876858 (accessed May 19, 2012); NEMA. 2007. General Cable acquires Chinese specialty cable facility. Press release, February 23. http://www.nema.org/media/ind/20070223b.cfm (accessed May 19, 2012).

40 GE Water. 2008. GE expands commitment to water technology in China with multi-million dollar investment. Press release, November 19. http://www.gewater.com/who_we_are/press_center/pr/11192008.jsp (accessed May 19, 2012); Toray Industries. 2009. Toray, China BlueStar establishes water treatment joint venture in China, to hold groundbreaking ceremony. Press release, August 25. http://www.toray.com/news/affil/nr090825.html (accessed May 19, 2012).

41 BP. 2011. The rising fortunes of the Caspian. *BP Magazine*. Issue 4. http://www.bp.com/sectiongenericarticle800.do?categoryId=9039889&conte ntId=7072885 (accessed May 19, 2012).

42 BP. 2005. Setting the standard. *Frontiers*. December. http://www.bp.com/liveassets/bp_internet/globalbp/globalbp_uk_english/ publications/frontiers/STAGING/local_assets/downloads/bpf14p22-31standardacg.pdf (accessed May 19, 2012). P. 26.

43 Economist Intelligence Unit. 2011. Economies of scale: How the oil and gas industry cuts costs through replication. *The Economist* report. http://www.managementthinking.eiu.com/sites/default/files/downloads/Oil%20and%20 Gas_%20Economies%20of%20Scale.pdf (accessed May 19, 2012).

44 SAP. [n.d.] Petronect: Reenergizing Latin America's leading oil and gas portal. Case study. http://www.google.com/url?sa=t&rct=j&q=&esrc=s&so urce=web&cd=1&ved=0CCUQFjAA&url=http%3A%2F%2Fdownload.sap. com%2Fdownload.epd%3Fcontext%3D27015EDBEAA93E8358A20 DBDE58CE7F82C5FBA779F76883FBBF76112C012AD82B0DDF32C786 801D3E997D5E8B6130DAEF1C1236A58CD9B26&ei=6Oc2T5CpFIPq2Q XIj8WBAg&usg=AFQjCNHLzF433HlXaaZIue8hfvKBd-U5wg&sig2=Et9R vNNo-oif1uHVpUsw1Q (accessed February 11, 2012).

45 Redden, Jim. 2011. Mud companies struggle with diminishing barite supplies. *World Oil Online*. December. http://www.worldoil.com/December-2011-Drilling-advances.aspx.

46 Ibid.

47 Hoch, Maureen. 2010. New estimate puts Gulf oil leak at 205 million gallons. *PBS Newshour*. August 2. http://www.pbs.org/newshour/rundown/2010/08/ new-estimate-puts-oil-leak-at-49-million-barrels.html.

48 Cavnar, Bob. 2010. *Disaster on the Horizon: High Stakes, High Risks, and the Story of behind the Deepwater Well Blowout*. White River Junction, VT: Chelsea Green Publishing.

49 EBARA Group. CSR Report 2011. Pp. 30–31.

50 Boston Strategies International press release. 2012. PTT group wins Boston Strategies International's 2012 Oil and Gas Award for Excellence in Supply Chain Management. April 25. http://www.bostonstrategies.com/images/ PTT_Group_Wins_BSI_Award_120425.pd.

51 Statoil. 2012. More fast-track developments on the Norwegian shelf. http://www.statoil.com/en/TechnologyInnovation/FieldDevelopment/ ONS2010ARealFastTrack/Pages/default.aspx/ (accessed June 3, 2012).

52 Flowserve. 2008. Supply chain: Does manufacturing have the capacity to meet projected demand? *Experience in Motion*. June. http://www.flowserve.com/files/www/Collections/Spotlight/Industries%20 Landing%20Page%20Spotlights/Power/EIM.pdf (accessed May 19, 2012).

53 ITT. 2006. Modular block bodies. http://www.duhig.com/Images/ITT/ pfmbb06.pdf (accessed May 19, 2012).

54 Thakkar, Shalini. 2010. BP's ties with Nalco Co.'s Corexit. *TopNews*. http:// topnews.co.uk/24532-bp-s-ties-nalco-co-s-corexit (accessed May 14, 2010).

55 Hitachi Cable. 2006 Annual Report. P. 18.

56 NSK. 2009. NSK reduces customers' costs with extended bearing life. Press release, January 28. http://www.4e534b.com/PressRelease/Jan282009.html (accessed May 19, 2012).

57 BP *Frontiers*, 2005.

58 Altus Logistics. [n.d.] Altus Oil & Gas moves the world's largest STP buoy. Case study. http://www.altuslogistics.com/i/Altus_case_FPSO.pdf (accessed May 19, 2012).

59 Saudi Aramco News. 2007. Program changing face of procurement. Aramco Expats. February 26, http://www.aramcoexpats.com/articles/ 2007/02/program-changing-face-of-procurement/ (accessed May 19, 2012).

60 Tenaris SA (TEN:BrsaItaliana). Stock quote & company profile. *Bloomberg BusinessWeek*. http://investing.businessweek.com/research/stocks/snapshot/ snapshot.asp?ticker=TEN:IM (accessed May 19, 2012).

61 2011. ADG signs $25m deal with Traverse (Daily Business Alerts [Australia]), May 28.

62 Wärtsilä. 2009. Wärtsilä and CEVA Logistics expand their cooperation. Press release, March 31. http://www.wartsila.com/en/press-releases/newsrelease302.

63 Price, Richard. 2006. Wood Group wins first supply chain services contract in Kazakhstan. Energyme.com, February 7. http://www.energyme.com/ energy/2006/en_06_0135.htm (accessed May 19, 2012).

64 Jacoby, David. 2011. RFID finds a "sweet spot" in offshore oil and gas drilling. *Logistics Digest*. January. Pp. 40–41.

65 Ibid.

66 Baker Hughes. 2008 Annual Report. http://files.shareholder.com/downloads/ BHI/1625622136x0x279889/ACAF6320-2A52-42B2-ACD9-B6A52C9A160A/ baker_hughes_2008_annual_report.pdf; Bloomberg. 2009. Air Products installs GE's revolutionary wireless monitoring. Press release, May 05. http:// www.bloomberg.com/apps/news?pid=newsarchive&sid=aXi45u6sq4UQ.

67 Jacoby, 2011.

6

MIDSTREAM—HYDROCARBON TRANSPORT EXAMPLES

Introduction: Supply Chain Cost Drivers and Relevant Design Constructs

Midstream environments vary in many respects, depending on the type of hydrocarbon concerned. An oil tanker operation has very different supply chain characteristics than an LNG pipeline operation or a tank farm. Accordingly, this chapter will provide high-level supply chain concepts that apply to all of these modes, and readers are further encouraged to seek specialist knowledge in their particular fields of practice. Many alternative transportation configurations are possible with the same exact inputs by changing the location, extent, and type of processing.

Petroleum that will eventually be used to make plastic can be transported and stored in crude oil, refined oil, or in petrochemical form. This means that midstream logistics can take advantage of *postponement*, or shipping in the lowest form of conversion, to lower inventory holding costs and buffer against potential drops in demand.

Natural gas can be transported and stored as

- LNG
- Compressed natural gas (CNG)
- Gas to solids (GTS)
- Gas to power (GTP)—transmitted to where it will be used as electricity

- Gas to liquids (GTL)—converted to methanol or ammonia for transportation fuel

- Gas to commodity (GTC)—used to produce aluminum or some other commodity that is energy intensive, rather than shipping it

Each option involves not only a transportation or storage cost comparison but also a different set of costs along the entire value chain. For example, transport in methanol form allows shipment in a conventional tanker but decreases the energy potential of the shipment substantially compared to other forms of gas, because methanol contains much less energy per shipload. In fact, each midstream option represents a completely different supply chain. Furthermore, owing to the specialized nature of midstream operations, specialized skills and knowledge in areas such as fluid mechanics and cryogenics (or at least enough background to understand the physical constraints) are required in order to make effective midstream supply chain decisions.

Furthermore, technological developments are changing shipping and storage options and their relative costs and consequently are making new supply chain opportunities available. For example, in the area of CNG, Enersea Transport's Votrans shipping technology and Coselle technology reduce gas losses for large quantities. For GTS, developments in natural gas hydrates are changing the economics of production, transportation, and regasification.[1]

Capital costs generally far outweigh operating costs. For LNG, laying a feed gas pipeline to a liquefaction plant costs $2–$5 million per mile, which encourages siting the liquefaction plant close to the gas field. Moreover, raw material and construction costs increased massively from 2005 to 2010, even after accounting for the drop and subsequent rise due to the global financial crisis. In contrast, liquefaction operating costs comprise only 2%–5% of the capital cost.[2]

This chapter explores unique aspects of supply chain management as it applies to midstream operations given these basic supply chain cost drivers. The key messages are as follows:

- In some cases it is possible to lower costs by altering the location, extent, and type of processing (changing the temperature, form, or lot size may reduce shipping costs).

- The financial and regulatory risks are so large and time frames are so long that it is best to ensure that all commitments to supply chain partners are based on final, approved decisions.

- The time frames are so long that solid financial analysis and hedging are necessary practices.

- There is a large learning curve, which favors using established supply chain structures, partners, and systems.

- Flexible production planning and control, with embedded safety and contingency plans, is essential to safety and profitability.

Project Risk Mitigation

In both oil and gas, midstream facilities are growing in number, size, and complexity. Today LNG distribution spans a global trade network, as shown in figure 6–1. The following examples illustrate the magnitude and visibility of midstream projects:

- KNPC's Mina Al-Ahmadi floating, dockside GasPort is the Middle East's first LNG receiving facility. The offshore aspect reduced capital costs compared to onshore LNG facilities owing to the absence of most conventional onshore real estate issues. Since the system is close to gas processing facilities and to the Kuwait gas grid, logistics costs will be low.

- Koch Pipeline and NuStar Logistics agreed on a pipeline connection and capacity agreement that will assure Koch Pipeline of future additional capacity to transport a growing amount of shale oil from the Eagle Ford shale to Corpus Christi, Texas, refineries and terminals.[3]

- Motiva Enterprises built an industry-leading biofuel (ethanol) distribution network. Motiva's supply chain includes rail and marine terminals, segregation and blending infrastructure, and export facilities, as well as partnerships with ethanol producers, railroads, terminal operators, and industry peers. This allows the company to handle ethanol for other major oil companies and fuel ethanol suppliers at a lower cost and with more flexibility than if each company used its own facilities, which is helping the industry comply with U.S. Renewable Fuel Standard regulatory requirements.[4]

- MarkWest Liberty Midstream will collaborate with Sunoco Logistics to build a 50,000-bpd ethane pipeline from the U.S. Marcellus shale to refrigerated storage facilities along the East Coast and then move them via ship to Gulf Coast markets.[5]

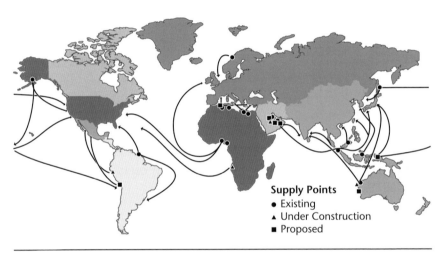

Fig. 6–1. Major LNG and oil tanker trade routes (*Source:* Shively, Ferrare, and Petty, 2010, p. 111.)

The public visibility as well as the long time frame of these projects brings huge regulatory risks and concerns, such as *NIMBY* ("not in my backyard"). Therefore, owners would be wise to secure a FID before committing to suppliers and to be aware of pending regulatory changes or upcoming revisions to standards. For example,

- The United States enacted a pipeline safety bill after a 2010 pipeline explosion in California killed eight people, requiring that operators install automatic shutoff valves on new or replaced pipelines and test the pressure of pipelines in populated areas.

- ISO 28460:2010 was instituted, which specifies requirements for on-loading and off-loading procedures and safe transit of LNG carriers through port areas; this standard applies to ship, terminal, and port service providers.[6]

- ISO 21809-5:2010 "specifies requirements for qualification, application, testing, and handling of materials" for reinforced concrete coating for oil and gas pipelines of concrete thicknesses of 25 millimeters or greater.[7]

- ISO, through its ISO/TC67 standard, is attempting to form an internationally accepted standard that incorporates the best aspects of many countries' individual standards. ISO 13623 (Pipeline Transportation Systems) is the general standard for pipelines and the governing standard of this committee; in addition, the group is working on standardizing substandards related to linepipe, coatings, cathodic protection, mechanical fittings, pipeline valves, and actuators. Working-group committees are composed of country-nominated experts in each subspecialty.

Visible evidence of safe and diligent operations can help to manage NIMBY concerns, public scrutiny, and potential delays due to regulatory oversight.

Engineering and Procurement of Equipment and Services at Minimum Total Cost and Risk

Because of generally long project time frames and high market risk, contracting methods and decisions are particularly important. Solid financial analysis is a requirement, and the use of financial hedging instruments is necessary in some cases.

Midstream projects are often cofunded by consortium partners owing to their size. The major shareholder often plays a key role in project management and/or selection of the EPC firm that manages subsuppliers. For example, a gas processing plant in Qatar leveraged a shareholder (Mitsui) as an EPC firm, which subcontracted to Hyundai for gas turbines, steam turbines, heat recovery steam generators, and desalination units. The project cost $2.95 billion, $20 million below budget. Another LNG processing project was established as a joint venture between a U.S. EPC firm and a Japanese EPC firm and involved two local joint venture partners to satisfy local content requirements. The contract was set up as a management contract with reimbursement for material costs plus an adder (profit margin), with a cap. While hundreds of suppliers were considered, only suppliers that demonstrated quality were allowed to bid. Low-cost bids from Asian suppliers drove the bids down noticeably.

At the next level in the supply chain, there are frequently bottlenecks caused by capacity constraints for key equipment. For LNG, compressors require significant lead time. Most suppliers follow a first-come/first-served prioritization of their capacity, but some have offered capacity reservations in exchange for firm orders or a certain volume of business, and others have sold production slots for cash.

Contracting methods

Buyers may be willing to pay a commitment fee for an option on capacity. Figure 6–2 shows how one owner used real options to decide whether to invest in a floating gas storage and off-loading unit (FGSO).[8] The company evaluated four substantive choices: (1) acquire options on steel to be able to build the ship; (2) acquire the ship itself; (3) refit a shipyard to construct the ship and possibly others; and (4) construct an offshore development in addition to the ship. Each option had a cost of acquisition and a cost of disposal or decommissioning if the choice turned out to be below expectations.

Considering the option value, the lowest-cost option turned out to be to refit the FGSO shipyard (fig. 6–3).

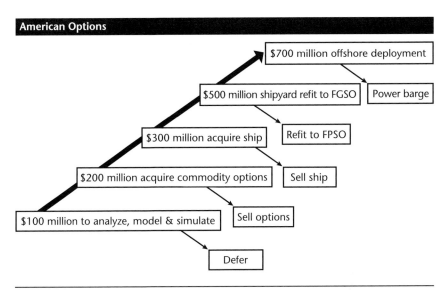

Fig. 6–2. Analysis of strategic investment in an FGSO hub using real options (*Source:* Anderson et al., 2008, p. 275.)

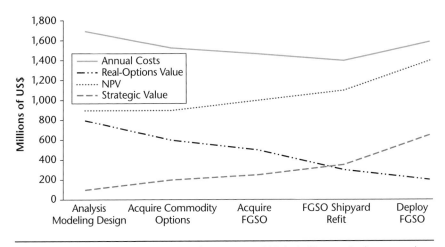

Fig. 6–3. Results of real options analysis on an FGSO hub (*Source:* Anderson et al., 2008, p. 275.)

Engineering and procurement

Standardization of design across multiple terminals can reduce capex significantly. At first it may seem that each project is unique—that there would be no learning from project to project—on the basis of the average size and large capex. Contracts are long (e.g., 20 years) and consortium models add complexity to contracting. However, there is a strong learning-curve benefit because contractors construct similar projects; thus, repeating the supply chain structure (suppliers, specifications, and systems) can take advantage of learning and reduce cost.

BG Group, which handles natural gas distribution, claims to operate the lowest-cost LNG terminals among its peers owing to the use of standardized designs. BG's Atlantic and Egyptian LNG terminals took only 4–6 years to build, including the tender to an EPC firm and the EPC tender to subcontractors. In contrast, competing LNG terminals at Rasgas, Qatargas, Oman (BG), and Nigeria (a consortium with BG participation) took 6–16 years to build, with the difference in speed resulting from the lack of a standardized design. BG Group's terminals operate at $175–$275 per tons per annum (tpa) (excluding financing and based on a two-train build), whereas the facilities at Rasgas, Nigeria, and Qatargas operate at approximately $300/tpa, $350/tpa, and $425/tpa. The difference in cost is also a result of BG's standardized approach. The Egyptian LNG project is owned by BG, Petronas, the Egyptian General Petroleum Corporation, EGAS (the Egyptian Natural Gas Holding Company), and Gaz de France. Its standardized design allows the consortium to lower capex to levels achieved only with smaller trains (3.6 million tpa), minimize design changes, and achieve more flexibility by having multiple smaller units, rather than a single large unit. The group claims lower purchase cost, shorter construction schedule, less financial and technical risk, and lower operating costs. On top of all those benefits, the project team claims higher reliability (96%) and availability (98.5%) than on other similar installations.[9]

The findings give hope for the many other offshore projects that are under way. No matter how large or deep, each one consists of components like wet and dry trees, jumpers, manifolds, risers, and supervisory control and data acquisition (SCADA) systems, which should be eligible for economies of scale and learning-curve benefits.

Construction and installation

Disciplined project management is a critical success factor for large midstream projects, since delivery and operational logistics need to be thoroughly planned in advance.

Nord Stream's pipeline provides an example of rigorous management of a physically challenging and complex project. The pipeline is being laid on time and within budget despite the large number of concrete pipe sections and the weight of each one.[10] The size of the project makes it a particular logistics challenge. Twin pipes will carry 55 billion cubic meters of natural gas per year over 1,224 kilometers, from Vyborg, Russia (near Russia's border with Finland) to Griefswald, Germany, via the Baltic Sea. Laying the pipe will cost $9.6 billion and involve 202,000 concrete-coated pipes, each of which is 12 meters long and weighs about 23 tons. Thirty pipe-laying vessels will work simultaneously on the project, including Saipem's Castoro Sei and Allseas' 300-meter-long Solitaire.[11] Galvotec Alloys, a supplier of anodes, used a complex, project-specific database to monitor the production and delivery of its anodes to the pipeline (delivery took up to eight weeks). The manufacturer produced over 13,000 pages of records to accompany the shipment.[12] Careful advance planning has helped to avoid surprises. In particular,

- Two logistics hubs—one in the Gulf of Finland, and one in Mukran, on the German island of Rügen—coordinated the physical delivery and materials management aspects.

- Time buffers minimized the risk of bottleneck activities, which has kept the project on schedule.

- Critical-path sections were executed four weeks before stipulated by the project plan, which enabled the rest of the project to continue according to schedule.

- External auditing and certification ensured confidence and reliability. Det Norske Veritas certified Galvotec's manufacturing and quality process.

In addition, at the equipment and component level, global sourcing is a key lever for keeping cost down. At the equipment and component level in the supply chain, global sourcing is keeping down the costs of compressors, blowers, and pumps. For example,

Kobelco bought 44% of China-based Wuxi Kobe Steel and will provide its process gas compressors intellectual property to Wuxi so that Wuxi can manufacture compressor systems in China at lower cost than Kobelco can elsewhere.[13] Many other turbocompressor manufacturers have also established joint ventures and made acquisitions in Asia to keep their costs down and tap into growing local markets.

Operations and Maintenance Cost Reduction

Key drivers of midstream operating and maintenance costs include the following:

- Production planning and control
- Maintenance of storage tanks and pipelines
- Driver efficiency

In pipeline operations, careful production planning and control makes a difference in throughput and eventually profit. A scheduler sends batch orders in a repeating sequence to avoid commingling fluids and to separate the ones that have less tolerance for contamination. On the basis of orders, or *nominations*, the scheduler assigns batches to available pipeline capacity—for example, a repeating cycle of diesel, gasoline, and jet fuel. Careful scheduling optimizes the delivery of the most profitable product to the customers in the time frame that they want it in, especially if there is a wide variety of products and grades.[14]

Reliability-centered maintenance and Six Sigma are important tenets of safe and precautionary maintenance in midstream operations. (These concepts have been covered in detail in Part 1.)

Tank storage is a key link in the midstream supply chain. Insufficient capacity causes logistical diversions to other facilities that do have adequate capacity. In addition, excessive or unplanned repair and maintenance increase operating costs. Corrosion is a perennial problem that must be addressed for safe operations. Tank storage is sensitive to safety and environmental concerns, whether the tanks are atmospheric, low pressure, or high pressure and whether they are single or double walled. Choice of materials (carbon steel, aluminum, stainless steel, or fiberglass-reinforced plastic [FRP]) can significantly reduce the corrosion problem. However, in most cases, corrosion still causes some degree of stress cracking, fatigue, galvanic corrosion, crevice corrosion, cavitation damage, and hydrogen embrittlement—just to name a few potential related problems. Corrosion can be addressed through engineering and maintenance practices related to coatings, as well as design changes to eliminate the source of vapor escape or other undesired contact, and to improve allowances, seal welds, and caulking attachments. Various API standards (e.g., API RP 570) apply, to ensure that maintenance operations protect leaks through impermeable barriers, volumetric measurement of level and mass, acoustic emissions, soil vapor monitoring, and variance analysis.[15]

Since fuel is a major operating cost, the choice of driver for pipeline compressors can have a significant effect on overall operating costs. Moreover, the rapid pace of technological change is increasing the efficiency of all of the available technologies over time. The choice of which driver to use—gas turbine, steam turbine, electric motor, or even diesel generator—determines the cost performance at the desired pressure and flow characteristics, given the types and cost of fuel available at the locations where it is needed. Next, upgrade and retrofit options must be periodically evaluated to increase the efficiency of older units as much as is practical. Motors made to comply with relatively new efficiency standards (e.g. NEMA Premium and IE3) can make for economically attractive upgrades. For turbines, the materials used in rotors and blades, as well as the machining processes used to make them, have greatly increased heat tolerance—and consequently, power generating efficiency. (More examples of power efficiency improvements through technology enhancements and upgrades are given in chap. 8.)

Inventory and warehousing management

Information technology is critical to advances in midstream productivity, helping operators know where the product is and offering more complete value-added solutions (e.g., consignment and vendor-managed inventory). Specific technologies in use include the following:

- SCADA. SCADA systems monitor pressure and warn of abnormal conditions. For example, PetroChina uses OASyS, a SCADA system, to monitor the pressure within the 2,100-kilometer Lan-Zheng-Chang oil pipeline, which has a total elevation change of 2,500 meters and passes through complex, varied terrain that can affect flow rates. The system provides early indications of leaks and potential disruption, thereby minimizing the negative impact of disruptions.

- Unmanned aerial vehicles. These have become cost-effective for pipeline surveillance in recent years.

- RFID. RFID tags were at one time problematic in liquid environments because the signal did not transmit through liquid. However, advances have made RFID an option for some applications. One vendor, Container Technology, is testing passive RFID tags for use in liquid and metal environments where RFID technologies did not previously work.[16] The tags, which comply with EPC Class 1 Generation 2 standards, can be placed on 55-gallon drums with a read range of about 20 feet.

Oil transport management

A growing trade imbalance is driving more use of parcel tankers (compartmentalized ships with multiple holds), rather than full vessels. Parcel tankers can manage these imbalances better because they can handle dozens of different types of cargo on the same ship (one per hold). Thus, in event of a trade imbalance, they are more likely to be able to maintain full loads in both directions by handling smaller lots of more cargoes. A shift of volume east of the Suez Canal (in the Middle East, Korea, and China) is driving the increase in production and export of base oils in the Middle East, while production and consumption have stagnated and, in some cases, declined in Europe and America during the same period.[17]

In addition, environmental regulations are driving tanker costs higher. For example,

- More stainless steel or coatings will be required in order to handle certain commodities such as Group III lubricants.

- Environmental requirements to burn low-sulfur fuel (e.g., SECA [Sulfur Emission Control Areas] 0.1% requirements) will raise fuel costs.

- Environmental requirements to treat ballast water will necessitate that carriers purchase and install chemical treatment systems and products, thereby incurring capex and raising operating costs.

- Because of negative publicity, major oil companies are increasingly reluctant to operate old ships that could be perceived as unsafe (even if they are safe). The drive to replace them with newer ships will raise costs.

Finally, deteriorating port infrastructure and piracy are driving midstream oil transport costs up.

- Lagging development of port infrastructure will force costs higher as vessels wait for access to ports. Once they are granted access, ports are shortening the time windows that the vessels are allowed to stay in the harbor, thus requiring faster on- and off-loading, which again costs more money.

- Piracy is a huge, costly, and ongoing threat. Because there is often little choice on routing, shipping lines are forced to use patrol boats to escort ships. Carriers must buy war risk insurance and pay crews bonuses owing to these security hazards.

LNG transport management

Most of the economics of LNG shipping is driven by LNG spot market prices. From a supply chain point of view, the key decision is whether to build a fleet. Today the existing fleet of LNG transport vessels is mostly privately owned.

The LNG tanker market experienced cyclical overcapacity crises in 1977 and 1981. Anglo-Dutch ships went unused and were sold and resold at successively steeper discounts. El Paso Energy ended up writing off its entire investment in LNG, completely scrapping three vessels. In contrast to the crude oil tanker fleet, the LNG fleet is privately owned and cannot be chartered, which limits the speed at which the industry can adapt to overcapacity.

Offshore facilities are becoming more prevalent for reasons of safety and economics. This may make onshore liquefaction and regasification terminals less desirable and eventually even obsolete.

Despite threats to the economics of LNG transport, LNG buyers are acquiring ships and terminals, and third parties are creating an independent market for regasification and shipping services, mostly on the premise of a much more flexible and dynamic market for LNG shipping and associated services in the future.

In summary, the LNG transport market is growing and changing rapidly. Most likely it will look very different in the years to come.[18]

Notes

1 Mokhatab, Saeid, William A. Poe, and John G. Speight. 2006. *Handbook of Natural Gas Transmission and Processing*. Burlington, MA: Elsevier.

2 Ibid.

3 Cantrell, Meredith. 2010. Midstream operators ramp up in the Eagle Ford shale. *Pipeline and Gas Technology Magazine*. November/December. P. 14.

4 Boston Strategies International. 2011. Motiva Enterprises LLC wins Boston Strategies International's 2011 Oil and Gas Award for Excellence in Supply Chain Management. Press release, September 26. http://www.bostonstrategy.com/images/Motiva_Wins_2011_BSI_Supply_Chain_Award_110927_dj.pdf.

5 Business Wire. 2010. MarkWest Liberty Midstream & Resources and Sunoco Logistics announce new Marcellus ethane pipeline and marine project. BusinessWire. June 1 http://www.businesswire.com/news/home/20100601007278/en/ MarkWest-Liberty-Midstream-Resources-Sunoco-Logistics-Announce

6 International Organization for Standardization. ISO 28460:2010. Petroleum and natural gas industries—installation and equipment for liquefied natural gas—ship-to-shore interface and port operations. http://www.iso.org/iso/ iso_catalogue/catalogue_tc/catalogue_detail.htm?csnumber=44712 (accessed May 19, 2012).

7 ISO. ISO 21809-5:2010. Petroleum and natural gas industries—external coatings for buried or submerged pipelines used in pipeline transportation systems. Part 5: External concrete coatings. http://www.iso.org/iso/iso_catalogue/catalogue_tc/catalogue_detail. htm?csnumber=40881 (accessed May 19, 2012).

8 Anderson, Roger N., Albert Boulanger, John A. Johnson, and Arthur Kressner. 2008. *Computer-Aided Lean Management for the Energy Industry*. Tulsa, OK: PennWell. P. 275.

9 Reddin, Phil, Rick Hernandez, Wesley R. Qualls, and Amos Avidan. 2005. Egyptian LNG: The Value of Standardization. Gastech.

http://lnglicensing.conocophillips.com/EN/publications/documents/ GastechValueofStandardizationPaper.pdf (accessed May 19, 2012).

10 Froley, Alex. 2011. Russian Nord Stream gas flows to Europe commence. Platts. November 8. http://www.platts.com/NewsFeature/2011/NordStream/ index (accessed May 19, 2012).

11 Nord Stream. 2010. Pipeline construction progresses on schedule. *eFacts*. December 16. http://www.nord-stream.com/press-info/emagazine/pipeline-construction-progresses-on-schedule-24/ (accessed May 19, 2012).

12 Garza, Rogelio E. and James Lenar. 2011. Sacrificial anodes selected for Nord Stream system. *Offshore Magazine*. May 1, http://www.offshore-mag.com/ articles/print/volume-71/issue-5/flowlines-__pipelines/sacrificial-anodes-selected-for-nord-stream-system.html (accessed May 19, 2012).

13 SteelGuru. 2011. Kobe Steel acquires 44pct stake in Wuxi Compressor Co Ltd. SteelGuru. May 18. http://www.steelguru.com/chinese_news/Kobe_Steel_acquires_44pct_stake_in_Wuxi_Compressor_Co_Ltd/205651.html.

14 Miesner, Thomas O., and William L. Leffler. 2006. *Oil & Gas Pipelines in Nontechnical Language*. Tulsa, OK: PennWell.

15 Myers, Philip. 1997. *Aboveground Storage Tanks*. New York: McGraw-Hill. P. 90.

16 Bacheldor, Beth. 2007. Manufacturer tests RFID to track industrial-size containers of liquid. *RFID Journal*. March 19. http://www.rfidjournal.com/ article/view/3156.

17 For more detail consult Soffree, Ronald. 2011. The shipping of base oils to/ from the Middle East: trends and challenges. Paper presented at the ICIS Lubricants Conference, Dubai, Oct. 12.

18 Tusiani, Michael D., and Gordon Shearer. 2007. *LNG: A Nontechnical Guide*. Tulsa, OK: PennWell.

7

DOWNSTREAM OIL AND GAS EXAMPLES

Introduction: Supply Chain Cost Drivers

Most types of refining are continuous processes with unique operations management characteristics related to the cost of shutdowns, turnarounds, and maintenance. This applies to distillation (separation of crude); conversion (decomposition, including cracking [or breaking down large molecules], unification [or aggregating small molecules], or reforming [rearranging molecules through isomerization]) or catalytic reforming; and treatment (desalting, hydro-desulfurization, solvent refining, sweetening, solvent extraction, and dewaxing). Blending, light-ends recovery, sour-water stripping, wastewater treatment, acid and tail gas treatment, and sulfur recovery are batch processes, which require different replenishment signaling, production scheduling, and materials management processes.[1]

The option to produce so many different products from the same input makes supply chain planning more complex in downstream oil and gas (and in petrochemicals) than in simpler process industries such as paper milling and sugar refining. A refinery can produce dozens of different products out of benzene, whereas in the minerals industry, a processing plant might be able to convert limestone into only two or three calcium carbonate products. To add to the complexity, some hydrocarbon by-products can be reused as intermediates in other processes (e.g., an olefins plant can produce aromatic by-products that can be recycled for use in fuels blending).

Investment costs are often roughly 10 times annual operating costs. For example, a plant with an investment cost of $2.8 billion might have annual operating costs (excluding raw materials and depreciation) of $314 million. Also, the choice of crude (crude oil evaluation relative to a reference crude) accounts for 70%–80% of the operating costs. The main operating cost is usually maintenance, accounting for well over half of total annual operating costs. Other significant operating costs include staff, power, and insurance.[2]

Therefore, refining economics are largely driven by the choices made during the planning phase, especially the choice of crude, processing routes (the number and type of processing units), and facility design. Operational productivity is largely limited by the physical design of the facility. Accordingly, this chapter stresses the following key messages:

- Project risk mitigation depends mostly on upfront capital investment decisions rather than operating costs.

- Single sourcing has been demonstrated to keep refinery projects on schedule by reducing complexity.

- Construction schedule adherence can be improved by ordering materials and components at the optimal level of bundling to make sure that all the materials, components, and whole systems come together at just the right time.

- Production planning can be optimized by (1) sequencing with fixed capacity constraints at the single-plant level and (2) modeling at the multiplant level.

- Access to specialized (and sometimes sole source or exotic) materials can best be ensured by establishing long-term contracts and deep research capabilities.

- Process control can optimize input grades, quantities, timing, capture of intermediates, and recovery of energy and waste to improve profitability substantially.

- Trade of refined product has become a global business, whereas blending, mixing, and storage is often a regional operation involving specialized facilities.

- Supply chain partners and contractors should be included in HSE planning, to ensure safety.

Project Risk Mitigation through Effective Capacity Management

Because of the aforementioned economic parameters, project risk mitigation depends mostly on limiting upfront capital investment risk, rather than on reducing ongoing operating costs. The wide varieties of refineries and crude oils (every crude oil is different, with some being paraffinic, while others are naphthenic, aromatic, or asphaltic) makes it hard to provide a set of guidelines on how to best manage supply chain costs and risks at a general level. The United States has mostly (66% of the total) coking refineries of three subtypes and some (29%) cracking refineries, with the rest (6%) being simple refineries (6%)—that is, plants that only perform atmospheric distillation, catalytic reforming, and refining with no further cracking or processing steps.[3] Most investment projects center around residuum processing options such as delayed coking and visbreaking or fluid catalytic cracking.

Capital investment risk in downstream operations can be mitigated through debottlenecking, upgrading process technologies, using more effective refinery catalysts, and optimizing production network design.

Some investment risk can be avoided by limiting the need for capacity expansion by debottlenecking existing plants. For example, ExxonMobil Chemical increased capacity for its Synesstic alkylated naphthalene base oil at its Edison, New Jersey, refinery by about 40% through a debottlenecking program.[4]

The need for refinery capital investment may also be limited by the deployment of new process technologies and catalysts that effectively increase the capacity of current plants. One industry-leading refiner recently conducted a multiyear study of new process technologies and refinery catalysts, to stay abreast of the new technologies.

Optimizing site locations and the refinery network may also reduce the need for overall capacity by ensuring the maximum economical shipping range from each plant. When designing supply chain networks, many applications exist for multi-echelon production optimization modeling. When considering where to place a new refinery, Pemex uses a wide range of criteria, including

- Type of technology project
- Configuration of plant
- Size of the new refinery
- Distance to sources of oil and consumer areas
- Availability of raw material
- Quality of crude available
- Operational efficiency with which it plans to operate the infrastructure
- Location, including environmental and social considerations
- Existing infrastructure (roads, pipelines, refining, etc.)
- Utilization of existing wastewater
- Cost of the land[5]

Engineering and Procurement of Equipment and Services at Minimum Cost and Risk

Limiting the number of suppliers involved in the management of capital projects can keep the projects on schedule by reducing interfaces. Furthermore, structuring of project management so that one party has control—either all or most of the way through—often improves project economics and timing.

Engineering and procurement/contracting methods

Single sourcing has been demonstrated in some cases to keep projects on schedule by reducing complexity. For example, DuPont builds, owns, operates and maintains sulfuric acid plants at its customers' sites to produce sulfuric acid from the output of the alkylation process and to convert refinery gases to finished products.[6] The value proposition to refiners is the solution's cost-effectiveness, the state-of-the-art technology, and DuPont's expertise in safe and reliable operation.

To illustrate the potential benefits of single sourcing, consider two case studies in the refining and petrochemical industry, with opposite sourcing strategies and opposite results:

- One operator awarded an LSTK contract to a sole-source supplier. The operator set the contract budget and negotiated the contractor price. Execution was on time (24 months), and the project met HSE goals. The close relationship between the operator and the EPC firm helped in ironing out problems with subsuppliers and transportation logistics.

- Another operator engaged an EPC firm to build a refinery based on a fixed price for labor and a cost-plus arrangement for materials. At the operator's direction, the EPC firm engaged many suppliers from thousands of prequalified bidders and hundreds of selected Tier 1 suppliers, and did not execute any alliance or partnering agreements. It also strongly encouraged low-cost country sourcing. This project lasted twice as long the LSTK project above and was not as profitable for the contractor.

Construction and installation

Before the design specifications are locked down, engineering can have a major influence over project price, by standardizing materials and equipment. Shell won Boston Strategies International's 2009 oil and gas supply chain award for its high-performing standardization program, which allowed it to reduce purchase prices by 30% for valves and cut variety by 50% through the use of its extensive *Materials and Equipment Standards and Code* catalog. The catalog, which is based on ISO and IEC standards to ensure interoperability, integrates 370 design and engineering practices (DEPs) that standardize tools and facilities, reducing recurring engineering and design work and consolidating spending on standard items. In addition to reducing the purchase cost, the practice also reduces delays due to supplier confusion and costs related to unnecessary rework. Continuous feedback from users and participation from external standards bodies keeps specifications up to date. Shell more than doubled the number of DEPs between 2000 and 2010 and aims for a DEP age of 3.5 years.

Life-cycle cost can also be reduced by specifying equipment that has lower operating costs over its active life (fig. 7–1). For pumps, the initial cost often represents less than 10% of life-cycle cost. Energy and maintenance costs are a much larger portion of total cost but are not always considered, often because of lack of time and lack of data.

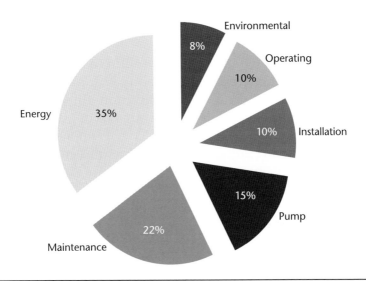

Fig. 7–1. Breakout of rotating-equipment life-cycle cost (excludes opportunity cost of downtime) (*Source:* Boston Strategies International, data adapted from International Energy Agency's 1998 Manufacturing Energy Consumption Survey.)

Construction schedule adherence can be improved by ordering materials and components at the optimal level of bundling. This is because many materials, components, and whole systems need to come together at just the right time. Refinery construction is a complex production plan involving materials, components, and whole systems that must arrive just in time—not early and not late. Major equipment, storage facilities, steam systems, cooling-water systems, and other systems must all be procured and delivered, built, or assembled on site. Rule-of-thumb estimates or cost curves for similar units constructed elsewhere with a similar capacity are used to predict costs. Appendix B contains a list of major and some minor equipment and major systems that may need to be procured or constructed. These include such major elements as the following:

- Coker
- Fluid catalytic cracker
- Hydrocracking unit
- Naphtha hydro-treater
- Reformer
- Isomerization unit
- Alkylation unit
- Polymerization unit
- H_2 unit
- Vapor recovery unit
- Gas plant
- Amine treater
- Shell Claus off-gas treating (SCOT) unit
- Pressure vessels
- Storage tanks
- Heat exchangers
- Piping system
- Valves
- Pumps
- Compressors
- Turbines
- Steam system
- Cooling-water system
- Desalters[7]

Operations and Maintenance Cost Reduction

Those who are responsible for delivering product to customers often divide planning and scheduling into three time frames, reflecting the degree of control over capacity at each time horizon. Strategic planning involves planning over the time frame that the capacity is fixed, which is usually at least several years. Master scheduling involves planning over the time frame during which demand is visible, which is often several months to a year or more. Production scheduling usually involves the time frame over which customers expect their orders to be delivered, which may be a few weeks to a few months. Whether for planning or execution, however, the problem is the same, so we treat all of the above requirements as *production planning*.

Depending on the scale of the refinery, on the number of grades and intermediate products, and on various other factors, production planning can involve intensive operations research models. Such models consider a large number of variables simultaneously (grade, cost, profit margin, batch size, by-products, intermediate processing, interplant transfer options, etc.).

Single-echelon production planning

Many refineries use spreadsheets with plans by tank. However, spreadsheets can reach their limits if varying grades of output have different profit margins and involve different processes, some of which are continuous and some of which are batch (e.g., blending and finishing). In these cases, single-echelon (i.e., single-refinery) production planning consists mostly of optimizing production sequencing within the constraint of a fixed plant capacity.

Large plants plan production with linear programs that consider the different blends of crude that are scheduled to arrive at different times to maximize profit on the basis of the characteristics of each crude (e.g., heavy, sour crude is more expensive to process), the profit margin of each blend, and the available capacity by unit.[8] Analysis can result in creative problem-solving and extraordinary productivity gains. One lubricants producer operates above nameplate capacity by vacuum pumping (increasing volume throughput without degrading quality), reducing viscosity, and shifting dewaxing between two refineries to optimize productivity. Despite these optimization efforts, the day-to-day work of dealing with planned variations (e.g., maintenance turnarounds and shutdowns to change product mix) and unplanned exceptions invariably complicates planning and scheduling.

Multi-echelon network planning

Multisite coordination is more complex than single-refinery planning, especially if the refineries produce intermediates that can be alternately produced at one or the other facility. In addition, the usual constraints—product quality, variable demand by finished product, and the cost of moving materials and finished product between facilities—factor into total cost. In a multiple-refinery configuration, the combinations of choices become so complex as to quickly require computational support and intelligence. Whereas crude oil is the primary input, intermediate outputs can include ethylene, benzene, chlorobenzene, and ethylbenzene, and final outputs can include products such as acrylonitrile, acetic acid, polyvinyl chloride, vinyl acetate, phenol, acetone, and acrylonitrile butadiene styrene.[9]

The problem of refinery optimization is more complex than that of optimization of traditional warehousing or distribution networking for discrete manufacturers. While the network design problem for a discrete products manufacturer may involve many SKUs, each SKU usually stays the same as it moves through the supply chain. Furthermore, in classic distribution warehousing problems, the demand for each product is independent, whereas in multiple-refinery production planning, demand for most intermediate materials and products is dependent on the next processing step.

Many refiners have adopted mixed integer linear programs to solve this application of a network design problem. For example, Unipetrol, a Czech refinery and petrochemical group, deployed AspenONE Advanced Process Control (APC) to improve profitability and reduce fuel consumption at its ethylene production plant. This solution increased the effective capacity of existing assets and cut $2.6 million a year in energy costs. Shell's lubricants marketing division also adopted AspenONE to optimize procurement of feedstock and maximize refinery profitability. The solution was integrated with a transportation and distribution network management application, which helped it respond more flexibly to market opportunities, and increase margin in a multisite environment. Multiplant and multilevel refinery optimization requires advanced modeling techniques. Khalid Al-Qahtani from Saudi Aramco and Ali Elkamel from the University of Waterloo have summarized the types of models that can be used to optimize networks of refineries. These include linear versus nonlinear, deterministic versus stochastic, static versus dynamic, and mechanistic versus empirical (empirical models are input-output models). Neural networks are a type of model that adjusts to patterns and relationships, which can be useful for modeling with variables related to mass and heat transfer, fluid mechanics, thermodynamics, and kinetics.[10]

In an effort to minimize network complexity, ExxonMobil consolidated its blending plants from 58 to 31 between 2004 and 2009. Concurrently the firm consolidated the number of customer order centers by 65% and rationalized its product range by 35%.[11] Similarly, based on network planning, BP Castrol decided to close its U.K. lubricant manufacturing plant and shift that work to Eastern European plants.[12]

Real-time process control

Real-time optimization of input grades, quantities, timing, capture of intermediates, and recovery of energy and waste can substantially improve profitability. This is often accomplished through automation software, sensors, analyzers, control valves, recorders, data acquisition and transmitting devices, and flowmeters.

Petrobras implemented AspenTech's process optimization solution at its ReVAP refinery in an equation-oriented real-time optimization initiative. Running nine times per day, the solution optimized feed selection, which effectively increased heavy crude processing capacity. Petrobras is evaluating moving to a multiple-refinery version of real-time optimization.

A major refiner optimized its fuel gas system to burn more efficiently and eliminate flaring. The solution involved a predictive controller embedded in Aspen DMCplus, along with multiple-refinery optimization and artificial intelligence. The project achieved significant energy savings while allowing the unit to maintain stable pressure and product quality.

Braskem implemented AspenONE APC software to identify operational changes that led to real-time optimization, thereby increasing ethylene production and maximizing product mix and overall profit margin. The system reduced energy costs and steam import rates.

Logistics: Transport, blending, mixing, and storage

As with upstream operations, transport of oversize and project cargo is characteristic of the logistics management challenges for downstream operations. For example,

- Petrobras used Gateway Logistics Group in 2009 to transport large pieces of equipment for the Abreu e Lima Refinery. Each piece weighed 119 tons and shipped from the Port of Houston.[13]

- Qatar Petroleum and Shell's Pearl GTL project awarded a shipping contract to the joint venture between Transcar and Transoceanic for the shipment of one million tons of construction supplies in 2006.

- For a single shipment, ConocoPhillips outsourced the hauling of two 300-ton coker drums from Kobe, Japan, to its refinery in Billings, Montana. The provider shipped the drums in two pieces each multimodally—first by ocean freight to Portland, Oregon, then by barge for 300 miles, then by 96-wheel tractor-trailer over 700 miles of back roads. The speed was limited to 35 miles per hour, and the truck haul was restricted to night hours, following local regulations. The planning took three years and resulted in a 700-page planning document.[14]

Blending, mixing, and storage is often a regional operation involving specialized facilities. If costs escalate, refiners often seek alternative sources from farther away. Several suppliers have revamped their distribution networks and associated services to become and to remain competitive sources of logistics services, often with the help of information systems. For example, Brenntag, a chemical distributor, bought Multisol Group to strengthen its ability to distribute lubricant additives and base oils in Europe and Africa and add to its blending and mixing capabilities.[15] Also, ExxonMobil implemented Invensys' InFusion-based system for inventory, order processing, packaging, and shipping operations. ExxonMobil has since expanded the application to other lubricant facilities.[16]

Disaster and environmental contingency planning

Refiners need to include their contractors and supply chain partners in HSE planning to ensure a safe and stable supply chain. Refineries experience more potentially fatal catastrophes than the next three industry sectors combined, according to OSHA, making safety a high priority with high stakes for failure.[17] In the wake of the 2005 explosion at BP's Texas City Refinery, which killed 15 workers and injured more than 170, OSHA inspected all of BP's U.S. refineries. As a result, the administration assessed nearly $3 million in fines at BP refineries in Ohio and Indiana for safety violations, in addition to the $21 million penalty imposed for the Texas City incident.[18]

New U.S. fatality reporting standards have made refinery accidents more visible, which heightens the likelihood of investigations and penalties. Through 2010, workplace fatalities were reported to OSHA only by a worker's employer; consequently, if a contractor was killed on the job, and the contractor was an outsourced service provider, the fatality may have been classified as occurring in another industry. In 2011, OSHA revised its record-keeping rules, requiring contractors to list the contracting agency when reporting any workplace fatalities.[19] The change in reporting requirements puts refiners at greater risk of OSHA inspections and financial penalties for safety violations.

In some cases, accidents may be blamed on outdated regulatory frameworks, so standards bodies are working to update these frameworks. Russia's current regulations governing the construction of refineries date back to the first half of the 20th century and remained practically unchanged since the mid-1960s. In 2011, GazpromNeft submitted a draft of Safety Regulations for Oil, Petrochemical, and Gas Processing Facilities to the Ministry of Energy.

In 2010, the EPA proposed amending the Clean Air Act to reduce greenhouse gas emissions from refineries and power plants.[20] The same year, API issued two new safety standards to help increase accountability:

- Recommended Practice 754, *Process Safety Performance Indicators for the Refining and Petrochemical Industries*, provides a set of process safety indicators for identifying events that may predict safety issues and for assessing the magnitude of those risks.

- Recommended Practice 755, *Fatigue Risk Management Systems for Personnel in the Refining and Petrochemical Industries*, provides guidelines to help reduce fatigue risk.

Nevertheless, accidents happen despite safety recommendations from institutions like API, indicating that operations management and supply chain management are essential instruments in ensuring safe operations. Tesoro's Anacortes oil refinery suffered an explosion in 2010, due to improper maintenance of a heat exchanger that subsequently burst, killing five people. The piece of equipment that blew apart was almost 40 years old. Prior to the accident, the welds that popped had not been inspected using a sophisticated method to detect cracks since 1998, according to an investigation by state regulators.

Huntsman, the chemical company, has demonstrated leadership in implementing an enterprise-wide operational risk application. The characteristics of this tool are instructive. First, it is integrative, meaning that it assembles risk information from a variety of frameworks and systems in the company. Second, it is Web based, which allows the project to involve employee participation to enter data into the system worldwide, invoking the principle of worker involvement (à la total productive maintenance and lean). Third, it is flexible, allowing for configuration and custom reporting. Fourth, it is to some degree predictive, and Huntsman expects the tool to capture knowledge about current and previous risks and apply that to future risks, to provide a predictive risk outlook.[21] To the extent that HSE and risk mitigation tools like this are used collaboratively across supply chains, supply chain risk will have been reduced.

Notes

1 Al-Qahtani, Khalid Y., and Ali Elkamel. 2010. *Planning and Integration of Refinery and Petrochemical Operations*. Hoboken, NJ: John Wiley & Sons.

2 Gary, James H., and Glenn E. Handwerk. 2001. *Petroleum Refining: Technology and Economics*. New York: CRC Press. Pp. 365–70.

3 Ibid.

4 *OEM/Lube News*. 2007. ExxonMobil Chemical completes debottleneck project increasing Synesstic (TM) alkylated naphthalene (AN) blendstocks production by 40 percent. *OEM/Lube News*. 3 (26; June 25). http://www.imakenews.com/lubritec/e_article000845700.cfm?x=b11,0,w (accessed May 19, 2012).

5 EBR staff writer. 2009. PEMEX selects site for $10 billion refinery. Energy Business Review, April 14. http://refiningandpetrochemicals.energy-business-review.com/news/pemex_selects_site_for_10_billion_refinery_090414 (accessed May 19, 2012).

6 PR Newswire. 2008. DuPont starts up clean technologies facilities to help refineries reduce their environmental footprints. PRNewswire, March 15. http://www.prnewswire.com/news-releases/dupont-starts-up-clean-technologies-facilities-to-help-refineries-reduce-their-environmental-footprints-56911657.html (accessed May 19, 2012).

7 Gary and Handwerk, 2001, p. 379.

8 Crude oil is defined by specific gravity (API rating), sulfur content (more sulfur makes oil "sour," and less sulfur makes oil "sweet"), and total acid number (TAN). For more on petroleum refining, see Gary and Handwerk, 2001.

9 Lidback, Alex. 2007. The ever changing conditions in the global aromatics market. Paper presented to the Procurement: Driving the Corporation into the 21st Century Spring Conference, ISM Chemicals Group, March 1; Al-Qahtani, Khalid Y., and Ali Elkamel. 2010. *Planning and Integration of Refinery and Petrochemical Operations*. Weinheim, Germany: Wiley-VCH. Pp. 4, 85.

10 Al-Qahtani and Elkamel, 2010.

11 ExxonMobil. *FD (Fair Disclosure) Wire*. 2009 Analyst Meeting. March 5, p. 23.

12 *OEM/Lube News*. 2009. BP Castrol to halt production at its Stanlow UK lube plant. *OEM/Lube News*. 5 (31; August 17). http://www.imakenews.com/lubritec/e_article001518003.cfm?x=b11,0,w (accessed May 19, 2012).

13 Gateway Logistics Group. 2009. Gateway Logistics Group acts as logistics provider for the first major equipment for the Abreu e Lima Refinery arrive in Pernambuco. Press release, October 2. http://www.gateway-group.com/en/releases/printview.asp?59 (accessed May 19, 2012).

14 ConocoPhillips. [n.d.] Publication CSH-11-0318 (brochure).

15 Brenntag. 2011. Brenntag completes acquisition of specialty chemical distributor Multisol Group Limited (UK). Press release, December 1. http://www.brenntag.com/en/pages/Presse/news/2011/MultisolClosing.html.

16 Market Wire. 2008. Invensys implements InFusion-based enterprise control system at ExxonMobil Lubricants Plant. Reuters, January 9. http://www.

reuters.com/article/2008/01/09/idUS154667+09-Jan-2008+MW20080109 (accessed May 19, 2012).

17 U.S. Occupational Safety and Health Administration. 2007. Directive CPL 03-00-004, June 7.

18 Fairfax, Richard. 2007. Statement of Richard Fairfax, Director of Enforcement Programs, Occupational Safety and Health Administration, before the Subcommittee on Oversight and Investigations, Committee on Energy and Commerce, U.S. House of Representatives. May 16.

19 OSHA. Standard 1904. 29 *Code of Federal Regulations*. http://www.osha. gov/pls/oshaweb/owasrch.search_form?p_doc_type=STANDARDS&p_toc_ level=1&p_keyvalue=1904 (accessed May 19, 2012).

20 Sinden, Amy, Rena Steinzor, Matthew Shudtz, James Goodwin, Yee Huang, and Lena Pons. [n.d.] Twelve crucial health, safety, and environmental regulations: Will the Obama Administration finish in time? University of Maryland Legal Studies research paper no. 2011-21. Washington, DC: Center for Progressive Reform. P. 10. http://papers.ssrn.com/sol3/Delivery.cfm/SSRN_ID1816731_code829721. pdf?abstractid=1816731&mirid=2 (accessed May 19, 2012).

21 Dyadem. 2010. Huntsman chooses Dyadem for operational risk management. Press release, September 29. http://www.dyadem.com/press-releases/ huntsman-chooses-dyadem-for-operational-risk-management (accessed May 19, 2012).

8

POWER INDUSTRY EXAMPLES

Introduction: Supply Chain Cost Drivers and Relevant Design Constructs

The cost of mistakes in the power industry is large, in terms of opportunity cost owing to high capital requirements, potential mismanagement during the construction phase (which can result in rework, penalties for not meeting delivery schedules, or operating inefficiency upon commissioning), and operating errors that can damage equipment or endanger personnel. The upfront investment required in order to build a power station starts in the billions of dollars. Making the wrong bet on technology, scale, or cost versus quality trade-offs can represent a major long-term headache.

Generating capacity typically comes in large chunks, with major gaps between increments. For example, U.S. utility Duke Energy faced the choice in 2006 whether to build two 800-MW coal-fired plants, and while it deliberated (18 months), the investment rose from $2 billion to $3 billion, mostly on raw material cost increases.[1] Nuclear power plants can cost twice as much.[2] The need for peak power further accentuates the investment and project management risk, magnifying any errors in the investment decision, especially if the plant operates at low capacity utilization.

These capital investment choices, and their resulting supply chain decisions, are further complicated by long lead times for critical path equipment, and choice of specifications that affect the plant's operating efficiency and maintenance costs for its entire life span. Turbine generators (gas, steam, or wind) typically have the longest lead time and are the most expensive single part of a power plant. However, ancillary systems from material handling to emissions control also affect operations and maintenance on an ongoing basis—for example, coal receipt and preparation, coal combustion and steam generation, environmental protection, condenser and feed-water system, and heat rejection (including the cooling tower).[3]

The complex interrelationships and trade-offs involved in setting up a large power-generating capability pose a conundrum

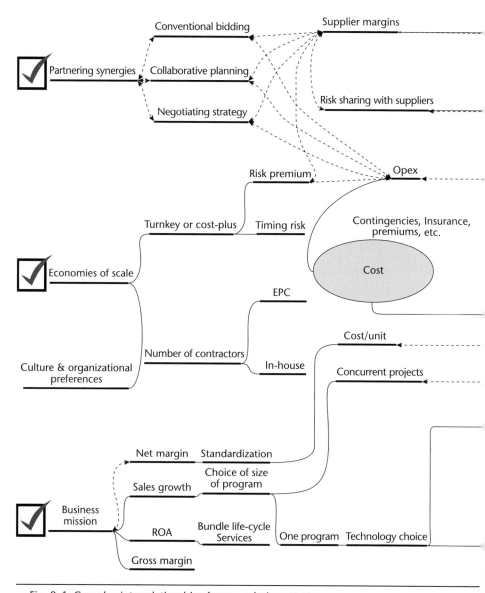

Fig. 8–1. Complex interrelationships favor a solutions strategy

for those in charge of major projects or programs. Supplier partnering strategies, technology choices, regulatory constraints, environmental objectives, and outsourcing decisions each involve uncertain costs and high risks. Moreover, these factors are mostly interdependent (e.g., changing the technology might change the outsourcing strategy, which could change the supplier partnering strategy), presenting a dizzying array of possible supply chain configurations for a major project (fig. 8–1).

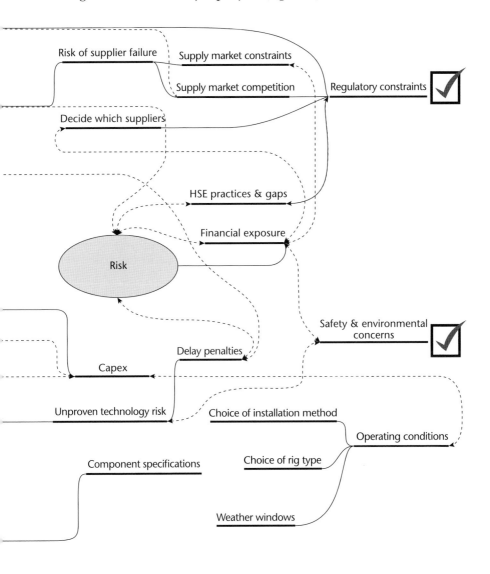

Project Risk Mitigation through Sequential Decision Making and Real Options

Financial and contractual exposure in large power projects is huge. Decisions must often be made pre-FID owing to long lead times for capital equipment. Formal government consent frequently does not occur until the project is long under way. Publicly visible supply chain failures are rare, but when they do happen, the consequences are grave.

Consider, for example, the failure of Greater Gabbard, in which Fluor Daniel absorbed losses of $343 million in 2010 when monopile foundation welding defects caused a project delay.[4] Subsea cable installation company Subocean went bankrupt because of the ordeal. The incident signaled a need for better financial control and risk reduction. Greater Gabbard is just one of many examples. One offshore wind operator had to deal with multiple suppliers going bankrupt when it took over responsibility for a project from its general contractor.

The most common approach is to throw data at the problem, resulting in an overload of data and projections, usually compiled at many levels with no framework for how to address risk at the rolled-up level. Traditional calculations yield a wide variation in results and are often divided into *high-medium-low* scenarios focused around a single indicator (e.g., NPV or levelized electricity cost); these scenarios do not quantify risk.

The best way to analyze risk—and therefore evaluate options for minimizing it—is to run robust scenarios to quantify the risks. One operator developed scenarios that each projected the cumulative effect over 30 years of standardization benefits, technology benefits, economies of scale, competition benefits, learning-curve benefits, and project management risks. The company compiled eight scenarios involving the aforementioned parameters that varied according to the number of suppliers, the type of technology, and whether an EPC firm was employed. It then valued each scenario and identified the lowest-cost option for the medium term and the lowest-cost option for the long term. (The lowest-cost option in the medium term was different than the lowest-cost option in the long term.)

Sometimes statistical and probabilistic tools, such as Monte Carlo simulation, are used to assess risks. However, several problems limit the effectiveness of such tools. First, they rely on critical assumptions that are often little or no better than guesses— for example, the shape of the probability distribution (normal, binomial, geometric, etc.) and the coefficients of likelihood of certain events (e.g., probability of supplier bankruptcy or technological failure). These *garbage-in/garbage-out* weaknesses frequently make probability simulation unlikely to be viewed with much seriousness by executives and boards of directors.

Once risks are analyzed and quantified, project owners need to decide how to best mitigate the risks. Off-loading risk is becoming an unavailable strategy, as EPC firms are no longer accepting the risk (or are charging a high premium to accept it).

Sequential decision-making is a relatively simple and still robust way to structure and manage risk. A sequential decision-making framework involves decisions along three of the most strategic supply chain decisions for large capital projects:

1. *Term*—time horizon for supplier commitments

2. *Bundling*—degree of bundling and solution buying

3. *Intimacy*—extent of collaboration with suppliers

For each dimension there is an associated matrix of options that highlights the degree of risk and of risk sharing that accompanies each option. For example, a flowchart helps to select the optimal form of project management, and a matrix helps to assess how much risk is involved in engaging one, two, and three suppliers for the major categories of expenditure.

A related risk management approach uses real options to build physical assets that can be expanded. This approach can keep as many options open as possible by *overbuilding* just enough to provide a platform for future expansion at minimal incremental cost. For example, a multipurpose installation vessel that costs more than a single-purpose one could be built in order to have the capacity for *any* type of installation, even though only one use (e.g., cable laying) is immediately anticipated. U.K. utility SSE created a real option by taking a 15% stake in a jackets supplier,

BiFab, in exchange for capacity up to 130 jackets per year plus an option on 50 more foundations per year for 10–12 years.[5]

Financial options can be created without physical assets, too. The first way is simply as an analytical construct—that is, "real-options" analysis can function like an NPV analysis. The Tennessee Valley Authority used real-options analysis to decide on power purchase options in 1994, and Enron used real-options analysis to make new-product development decisions, switching options for gas-fired turbines in 1994.[6]

Engineering and Procurement of Equipment and Services at Minimum Cost and Risk

Equipment standardization

Although facilities and equipment are highly engineered, a certain degree of standardization is possible and economical. Rolls-Royce pioneered the concept of modular design in 1970s to reduce the cost of customized parts and to simplify maintenance. The modular design concept helped Rolls-Royce achieve economies of scale in research and development, production, and inventory management. Based on the modular design concept, the Rolls-Royce RB211 gas generator is made of five modules that can each be replaced individually for maintenance. The units can also be retrofitted effectively by replacing only those modules that have been updated. Similarly, Rolls-Royce's RCB barrel compressor is based on a modular design wherein both stationary and rotating components can be removed simultaneously.[7]

More recently, motivated in part by long lead times for turbines, GE and Siemens have been standardizing their offerings by reducing engineering and design customization to reduce construction times. Siemens has also modularized steam turbine design. Siemens has twelve basic power plant combinations: four for simple-cycle gas turbines, six for combined-cycle plants, and two for coal-fired steam power plants. The modularization and design changes have the potential to reduce construction times and lower costs of construction and operation.[8]

The concept applies to pipes and valves as well. A U.S. electric and gas utility lowered the cost of its pipes, valves, and fittings through standardization. After reviewing more than 1,000 items across and 47 stocking locations, a cross-functional team of engineers, consultants, and procurement professionals identified 24% savings by standardization of medium-density polyethylene (MDPE) pipes. Many low-density polyethylene pipes were upgraded to MDPE to achieve the scale and volume purchasing leverage that suppliers would offer owing to standardization. The team also standardized on corrosion coatings according to a three-part matrix based on the application, pipe diameter, and operating company.

Solution bundling

A popular form of bundling for power utility capital projects is combined purchase and operating/maintenance agreements related to gas and steam turbines. Many combined agreements now include ancillary services such as commissioning, testing and calibration; user training and engineering support; operation; maintenance; and management of spare- and repair-parts inventories. The aircraft industry refers to such outsourced life-cycle agreements as *power-by-the-hour agreements*. For example, wind turbine suppliers also install foundations, but buyers can sign independent contracts with the foundation suppliers and the wind turbine suppliers if they wanted to weigh the cost complications savings against the risk of subsequent warranty erosion and possible operating and maintenance.

Instances of bundling are numerous, and the following examples will be instructive for those wishing to benchmark their supply configuration and performance:

- Rolls-Royce signed an eight-year contract with BP in 2007 to provide maintenance for the 28 RB211 turbines in BP's Azerbaijan plant.[9]
- Dresser Rand bought Leading Edge, a Houston-based gas turbine service company, for its repair and maintenance capabilities.[10]
- Toshiba and IHI formed a joint venture to provide maintenance services for steam turbines.[11]

- MAN Turbo bought a steam turbine repair company that does repairs and upgrades and makes spare parts for turbocompressors and steam and gas turbines.[12]
- Some wind turbine suppliers also install foundations (however, buyers can sign independent contracts with the foundation suppliers).
- Capstone Turbine developed a fixed-price maintenance program for scheduled and unscheduled maintenance.[13]

The decision whether to engage suppliers on a bundling basis is based partly on cost-effectiveness and partly on the strategic importance of repair and maintenance operations to the operator's business mission. Usually, operators feel that operations and maintenance is a core competency. Moreover, some want the capability not only for their own operations but also for extension to other power companies on a third-party basis.

Supply risk mitigation

Treating vendors as partners is one of the best forms of supply risk mitigation. Commenting on the importance of collaborative procurement, Peter Hessler, author of *Power Plant Construction Management*, explains,

> Although most company procedures require the typical vendor selection process to follow a "three quote and select the lowest bidder" scenario, that is exactly what often drives the relationship to be adversarial. An alternate approach is to use the bidding process only for identifying and prequalifying the suppliers. Then, the next step would be geared to maximizing value creation, as opposed to reducing costs by squeezing suppliers' margins and scope. However, entering into a search for mutual value creation requires an understanding of each party's objectives and finding ways to achieve fair resolutions to common issues.[14]

The following examples demonstrate how companies that use power equipment have effectively employed partnerships to strengthen their performance reliability and lower cost:

- ABB of Canada and Cominco, a U.K. mining and metals producer, mutually benefited from their partnership. Cominco decided to develop ABB as a supplier to new operations in South America, while ABB recommended a less costly solution than Cominco expected. The two companies developed measures to evaluate the partnership, including business process efficiency, life-cycle costs of products, and customer benefits from technical advances. The cooperative relationship is based on long-term mutual benefit, architected at strategic levels with executive sponsors and partnership managers, without a contract.[15]

- China Light & Power considers its suppliers to be an integral part of its business, so it seeks mutual benefits and shares its vision and goals with them as key business partners. The company adopted a risk-based supplier assessment system, which provides systematic performance feedback, to evaluate its major contractors and suppliers.[16]

- Mitsubishi Heavy Industries (MHI) formed a joint venture with Hangzhou Steam Turbine & Power Group. MHI will benefit from increased sales opportunities in China and lower labor costs. Chinese buyers will benefit from shorter lead times for high-technology equipment. MHI will export key components from Japan, to ensure quality and protect copyright.[17]

- Doosan helped its suppliers implement best practices, including initiatives traditionally covered under lean—such as collaborative planning, forecasting, and replenishment; simplification and standardization; and e-procurement. It will provide subsidies and other assistance to small and local suppliers. Doosan expects the initiative to reduce component lead times and cost.[18]

Vertical integration may seem like an easy way to ensure component supply but should be pursued only if the synergies are intended to be permanent. If a supply shortage appears temporary or short lived, one party would need to exit the relationship eventually. The following are examples of companies that have integrated vertically:

- Siemens took equity in Yangtze Delta Manufacturing and GIS Steel & Aluminum Products, two Chinese aluminum foundries, to lower costs and facilitate sales in China.[19]

- BHEL began manufacturing components such as nuclear heat exchangers that it had previously bought from external suppliers. It made the move to support a long-term projected increase in demand.[20]

- Dong acquired A2SEA, an installer of offshore wind farms, in 2009 (and Siemens took an ownership stake in the venture in 2010).[21]

- Toshiba bought into supplier Nuclear Fuel Industries to gain more ready access to nuclear fuels.[22]

To prepare for partnerships, power utilities must use a form of supplier qualification. The usual ratings and criteria apply, as discussed in Part 1. However, to identify suppliers with long-term potential that will ensure that they will be able to support a 20–40-year (or longer) project, an evaluation of the success factors for the current major players may be helpful. The top four suppliers of electrical equipment—Siemens, ABB, Schneider, and GE—have a combined market share equal to more than a third of the supply markets for major equipment of interest to electric utilities.[23] How did they get to be so strong? A study of their rise to prominence reveals three prominent success factors: (1) technological innovation and research investment, bringing high visibility projects and creating the market; (2) expansion into Asia; (3) entry into nuclear power and wind power.

Siemens is a leader in electrical distribution and control equipment, lighting, and motors. Its original company, Siemens & Halske, was founded in 1847, based on the invention of the telegraph. It followed a series of technological innovations in telecommunications. For example, in 1848, the company built the first long-distance telegraph line in Europe, and a decade later, it built telegraphs in Russia. It expanded into Asia starting in 1923, when it established a Japanese subsidiary. As early as 1969, Siemens had nuclear power activities under the umbrella of Kraftwerk Union.

ABB is a global leader in electrical distribution and control equipment, high-voltage cable, and motors, among other electrical supplies and equipment. It acquired much of its technology through acquisition. Created in 1988 by merging two existing electrical equipment suppliers (Swedish ASEA and Swiss Brown, Boveri, and CIE), ABB made 55 acquisitions in 1988 and 1989, one of which was Westinghouse Electric (the remnants of Westinghouse's power division after Siemens purchased its industrial gas turbine unit and Rolls-Royce purchased its Marine division). Westinghouse had an extraordinary amount of technology and intellectual capital, especially for nuclear power plants. In 1990, ABB expanded aggressively into Central and Eastern Europe. In 1992, it established more than 20 new manufacturing and service units in Asia, largely through joint ventures and acquisitions. ABB won several $90–100 million nuclear power plant contracts in the United States and Germany starting in 1999. It currently operates a wind turbine generator factory in Vadodara, India, and recently invested in a company that senses the wind flow in front of wind turbines so they can be aligned toward incoming breezes for maximum power output.

Schneider Electric is a worldwide leader in electrical distribution and control equipment and uninterruptible-power-supply systems, among others. Founded in 1836, it entered the emerging electricity market in 1891, producing electrical motors, electrical equipment for power stations, and electric locomotives. Its international acquisition spree began in 1988, with Telemecanique, and included Square D in 1991 and Merlin Gerin in 1992.

GE—a leader in many electrical equipment categories, as well as gas and steam turbines—got its technology start in 1893 when Thomas Edison and Henry Thompson bought Rudolph Eikemeyer's electrical business and it absorbed the National Electric Lamp Association in 1911. GE then introduced the first superchargers (turbines) during World War I and continued to develop them during the interwar period. It acquired renewable technology in 2002 when it bought the wind power assets of Enron during its bankruptcy proceedings. GE invested in engineering and supplies for the new wind division, and its sales doubled to $1.2 billion in 2003.

Because of the strength that the leaders possess in technology, operators have a lot to gain from technology partnerships. GE and Siemens have shown a propensity to enter such agreements. GE

Oil & Gas signed a contract manufacturing agreement with Triveni Group, an Indian steam turbine manufacturer. The joint venture, under which GE will transfer technology to Triveni, will produce 30–100-MW steam turbines and will allow GE to offer products at lower cost.[24] Siemens recently licensed a composite blade casting technology from Mikro Systems, one of a long string of such technology acquisitions. The technology also cuts lead time, reduces design and production costs, and raises turbine efficiency via increased temperature tolerance of the blade.[25]

In those supply markets where technology is critical and market concentration is high, buyers may not always have a wide choice of suppliers so should be aware of situations where suppliers become dominant to the point where they may influence prices or the pace of innovation. Under most international legal systems, certain types of dominance are illegal, such as price fixing (horizontal and vertical), price signaling, predatory pricing, price discrimination, promotional discrimination, and geographic discrimination (carving up markets). Each country has a trade commission that deals with concerns about market competitiveness.[26]

In a hallmark case from 1963, customers accused Westinghouse Electric of price signaling, a practice in which oligopolistic competitors share their pricing strategies with each other. The case, documented in a Harvard Business School study, explains how the cyclical and then-depressed market for turbine generators had dropped 50% from its peak and seven executives were convicted of conspiracy over pricing for various heavy electrical products. Customers sued GE and Westinghouse for over $100 million each. The case is about a pricing policy that appeared to be public and fair (because there was a pricing formula) but was found to be a form of price signaling, allowing them both to prop prices up without competitive pressure. Thus, while forming partnerships with leading companies in a concentrated industry, buyers need to establish checks and balances to ensure competitive pricing.

Long-term contracts are an excellent way to mitigate supply risk. For example, Parker Hannifin's aerospace group signed a 40-year contract to design and supply valves, oil pumps, and related parts for the Rolls-Royce Trent XWB engine. Consistent with the *service-technology-premium* paradigm for raising the value of manufactured products that was outlined in Part 1, Parker also designed the entire hydraulic system. Eaton signed a similar 40-year agreement with Rolls-Royce to design, develop, and supply parts.[27]

Substitute or alternative materials may reduce the risk of price increases and material availability. However, the strategies need to be well thought out. For example, using cement as an alternate material for steel as a way of mitigating price risk most likely would not make sense, since the two prices frequently move in the same direction.

Global sourcing—that is, looking outside one's own region for suppliers—has proved to be a good way to mitigate risk of material unavailability. Turbine manufacturers sought suppliers of forged and cast components in Asia after the recession, based on the perception that Western suppliers presented few options or ideas on how to reduce lead time whereas Asian suppliers were aggressive, nimble, and price competitive. Also, both Siemens and ABB engaged Stalprodukt, a Polish company, to ensure adequate supply of electrical steel that had been in short supply in the West.[28]

Construction and installation risk

Construction management of welders and other craftspeople is a large part of the job of project management during the installation phase, but from a supply chain management point of view, construction contract management is where financial and operational commitments can create or mitigate exposure, owing to supply chain relationships.

Terms and conditions of the contract can align (or drive apart) the interests of the owner and the construction contractor. Liquidated damages clauses for delayed completion are often subject to negotiation; a cap of 20% of contract value may serve as a benchmark. The need for extra work and delays are both subject to negotiation, depending on the situation and the possible causes of project extension, expansion, or delay. Certain other clauses tend to be more standard, including indemnification, liens for obligations not yet fulfilled, and termination for cause or convenience.[29]

Bonds ensure contractor fidelity and integrity. *Bid bonds* ensure that the contractor performs the work once awarded the contract. *Performance bonds* ensure that the contractor fulfills obligations once engaged. *Payment bonds* ensure that employees and subcontractors of a contractor are paid. Insurance coverage and insurance pass-throughs such as naming additional insured parties on the contractor's policy deserve careful attention to prevent ambiguity or lack of coverage in the event of incidents or accidents.

Unbundling construction and installation services to reduce cost is worth considering on a case-by-case basis. Sometimes a direct buy from a subcontractor can reduce unit cost; however, the burden of integration and the liability for faulty integration is then placed on the buyer. The question of how far to unbundle inevitably arises: If unbundling original equipment and service, why not also unbundle the major components of the original equipment? Buyers need to decide quickly what business they are in and target the degree of integration that fits their business model.

Operations and Maintenance Cost Reduction

Like any process manufacturing operation, production management and maintenance go largely unnoticed if all goes well but face severe scrutiny if systems fail to function effectively or if there are accidents. Both of these scenarios have the potential to be financially disastrous.

Production management

Production management in power plants is analogous to master scheduling and load planning in a factory. Long-range planning is handled by *block loading* total energy or load over a period of seasons or years. Fuel budgeting and operations planning is managed by *incremental loading* over a period of weeks or months. Staffing and power pooling plans are generated on a daily, hourly, or weekly basis. Simulations using probability can test what would happen under various loads, to estimate long-run production costs. They can also be used to determine how to get the most power out of the least amount of fuel, giving rise to a multitude of optimization techniques.

Short-run planning has traditionally been based on one of three production scheduling methods: (1) Operating at peak utilization and exporting power when it is not needed; (2) operating at low utilization rates and increasing output when necessary; or (3) operating at a constant rate.[30] Technologies are increasingly allowing variable rates of production that can meet demand or supply, with a dramatic impact on cost efficiency by leveling demand patterns and allowing variable production rates, both of which further alleviate the peaking problem. For example,

- Smart-grid technology and related technology has, to some degree, mitigated the peak demand challenge by improving the ability to efficiently distribute electricity over the transmission grid and over time.

- Capacitors and compact fluorescent lighting technologies smooth consumption.

- Demand-response programs price energy to encourage off-peak consumption.

- Variable-speed drives have made a difference in power plant operation, as in many other applications. ABB improved the energy-intensity index of a fluid catalytic cracking unit by 10.5% within one year.[31]

- Wind turbine sensor systems adjust the blade pitch and other operating parameters in response to changes in wind speed and direction.

Maintenance management

Reliability engineering is essential to keep uptime high. Involvement of personnel keeps a focus on root cause problem solving. Good root-cause analysis will result in a Pareto chart of the root causes of failure. In one case as an example, the principal causes were identified, in order of occurrence, as boiler repair, turbine repair, generator, feed water, controls, ash water, baghouse, blowing air, and circulating water (fig. 8–2).[32]

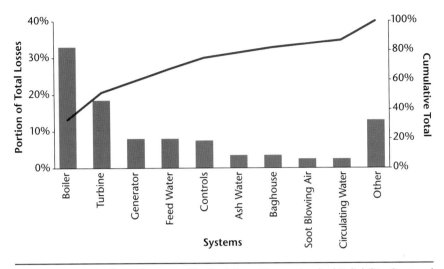

Fig. 8–2. Pareto chart of causes of boiler failure (*Source:* Applied Reliability-Centered Maintenance, August, 1999, p. 117.)

Life-cycle cost management is critical to optimizing the cost of rotating equipment. Unfortunately, life-cycle cost analysis is complex when dealing with turbine systems, and many utilities do not have good data or analysis tools. One turbine supplier has developed an economic valuation model that calculates the operating costs for two comparable units to make apparent how much one would cost versus the other under a multitude of specific operating conditions, applications, and parameters. It takes into account, for example, hours at base load, number of starts, and operating temperature. It computes an output that shows the amount of power produced, the efficiency rate, emissions rate, and uptime of each system.

RFID can help improve maintenance productivity and uptime in power plants. Power utility applications of RFID include the improved ability to

- Locate aging fixed assets (pipelines, between walls, underground, etc.)

- Distinguish one valve or piece of equipment from another similar one nearby, to avoid repairing or replacing the wrong equipment (RFID is used in medical equipment for a similar reason—namely, there have been unfortunate cases of major surgeries being performed on the wrong patients)

- Access repair status and history on an asset, to facilitate better diagnostics

- Manage assets, so expensive tools do not disappear (one turbine maintenance facility lost millions of dollars of equipment shortly after a layoff, as the laid-off employees may have taken the liberty of walking away with the tools)

- Track and manage inventories of spare parts

- Track employees

- Track vehicles

- Schedule maintenance

- Access management and infrastructure security through access/entry controls[33]

Predictive maintenance. Applied information technology is increasing the productivity of plant operations everywhere and is especially useful in predicting impending failures. Sensors and actuators play a big part in providing valuable information on impending performance degradations and therefore play a major role in increasing uptime and reducing cost.

Steam turbine sensors and actuators provide information on overheat and over-pressure situations and sometimes provide automated responses. When integrated with logical algorithms that can interpret patterns of information, they can provide better diagnostic information than manual operators can.

Steam turbine blade sensors can help avoid unnecessary preventive maintenance by detecting blade cracks without opening the case. For example, Monitran Technology developed a blade monitoring system that collects condition data and transmits it via a traffic light system. The information obviates the need for detailed turbine blade testing during maintenance.[34] GE is also applying intelligent condition monitoring to motors through a new Turkish acquisition called Artesis, which makes intelligent condition monitoring products designed to detect anomalies on various motor types, sizes, and loads.[35]

Inventory management

At the simplest level, utilities should pool maintenance, repair, and operating inventory across multiple companies. An electric and gas utility optimized inventory for its six operating companies, by conducting training in replenishment philosophies and mechanics, readjust demand triggers, and resetting reorder points. Through the effort, the utility identified and wrote off obsolete inventory, set inventory targets at the SKU level, and managed an inventory reduction program.

Vendor-managed inventory frequently results in lower inventory management and carrying costs. ABB worked with its own suppliers to establish updated delivery and stocking procedures, whereby the suppliers would manage the inventory instead of ABB doing it. Benefits included shorter lead times and stock reduction.[36]

Parts kitting can reduce field service costs by assembling frequently used kits in an assembly-line fashion by lower-cost employees in a standardized setting, as compared to having kits assembled by higher-cost field labor on an ad hoc basis.

Rapid inventory replenishment, even if expensive, can sometimes reduce costs by getting equipment up and running sooner. Weir implemented a rapid response program for hydraulic valve controls and experienced a 25% increase in orders in 12 months—proof that customers found the concept worthwhile.[37]

Performance-based logistics contracts can provide suppliers a strong incentive to ensure uptime and output targets are met. Some utilities are implementing bonus compensation whereby suppliers get paid more if capacity exceeds the nameplate amount—or less if it falls below. In a case of below-expectations performance, the Indian government purchased turbines that failed repeatedly at a high-visibility power project, and the ordeal devolved into a blame game. Careful negotiation of accountability for system uptime can improve overall system performance. There are many examples of performance based logistics contracts in the U.S. military. For example, the U.S. Army held GE to engine availability targets for its GE's T700 Engine Program, and GE used lean management to cut engine turnaround times from 265 days to 70 days.

Environmental contingency planning

For the most part, HSE standards in power generation are based on cross-industry standards and country-specific legislation, making it incumbent on the supply chain staff to understand general HSE standards, as well as local rules and regulations in the jurisdiction of power generation, and to ensure supplier compliance with the regulations. For example,

- In the United States, a number of government regulations encourage utilities to be prepared for crisis management in the event of disaster. These include the North American Electric Council's cybersecurity standards, which require power plants to continuously access and enhance their security systems; the Homeland Security Act of 2002, which identifies critical infrastructure such as utilities as high priority; and the Sarbanes-Oxley compliance, which deems utilities to be Level 1 critical systems, which are expected to be restored within 48 hours.

- In India, Tata Power's Haldia power plant implemented ISO 14001:2004 (environmental management system) and OHSAS 18001:2007 (health and safety management system).

- In Australia, the Shaw River power project implemented the Environment, Health, and Safety Management System (EHSMS) for the power station and corresponding gas pipeline. EHSMS is based on international standards such as ISO 14001:2004 and Australia/NewZealand standard AS/NZS 4801:2001 (occupational health and safety management).

The World Bank has established a worldwide standard, Environmental, Health, and Safety Guidelines for Electric Power Transmission and Distribution, which contain performance levels and measures that it considers achievable at new facilities by using existing technology and at reasonable cost. For example, the guidelines provide a table for minimum working distance for employees for different voltage levels (taken from OSHA 1910, subparts R and S). It also states that the guidelines are dependent on the host country and that if the measures of both differ, then the more stringent one is to be adopted (unless specific project circumstances dictate otherwise).

Notes

1 Wald, Matthew L. 2007. Price of new power plants rises sharply. *New York Times*. July 10. http://www.nytimes.com/2007/07/10/business/worldbusiness/10iht-power.4.6593271.html.

2 Rascoe, Ayesha. 2012. U.S. approves first new nuclear plant in a generation. Reuters. Rockville, Maryland, February 9, 2012. http://www.reuters.com/article/2012/02/09/us-usa-nuclear-nrc-idUSTRE8182J720120209.

3 Woodruff, Everett B., Herbert B. Lammers, and Thomas F. Lammers. 2004. *Steam Plant Operation*. New York: McGraw-Hill.

4 Collins, Leigh. 2011. Fluor learns costly EPC lessons from Greater Gabbard. Recharge News. April 1. http://www.rechargenews.com/energy/wind/article250629.ece.

5 Sharp, Tim. 2010. SSE buys 15% stake in BiFab. *Herald Scotland*. April 12. http://www.heraldscotland.com/business/corporate-sme/sse-buys-15-stake-in-bifab-1.1020234 (accessed May 19, 2012).

6 Copeland, Thomas E. 2001. *Real Options: A Practitioner's Guide*. Texere. New York: LLC.

7 Prencipe, Andrea. 1998. Modular design and complex product systems: Facts, promises and questions. CoPS Publication No. 47. http://www.cops.ac.uk/pdf/cpn47.pdf (accessed May 19, 2012); Rolls-Royce. RB211 Gas Turbines: For power generation and mechanical drive. http://www.google.com/url?sa=t&rct=j&q=&esrc=s&frm=1&source=web&cd=1&ved=0CGoQFjAA&url=http%3A%2F%2Fwww.rolls-royce.com%2FImages%2Fer200na_tcm92-21095.pdf&ei=giTMT8e8E8L00gHIo6B8&usg=AFQjCNEeE3mEXtvhRSoRMHUpw5wv3AvSvg (accessed February 8, 2012); Robb, Drew. 2011. Compressor maintenance trends: Modularity, remote monitoring and outsourcing are key. *Turbomachinery International Magazine*. September/October. http://www.turbomachinerymag.com/sub/2011/SeptOct2011-CoverStory.pdf (accessed November 8, 2011).

8 Soares, Claire. [n.d.] Gas turbines in simple cycle & combined cycle applications. National Energy Technology Laboratory. http://www.netl.doe.gov/technologies/coalpower/turbines/refshelf/handbook/1.1.pdf (accessed May 19, 2012).

9 Rolls-Royce. 2007. Rolls-Royce wins £120M TotalCare® contract for industrial gas turbines from BP. Press release, December 18, 2007. http://www.rolls-royce.com/energy/news/2007/rr_wins_totalcare.jsp (accessed May 19, 2012).

10 Dresser-Rand. 2010. Dresser-Rand acquires Leading Edge Turbine Technologies, Inc. Press release, January 18. http://investor.dresser-rand.com/releasedetail.cfm?releaseid=438238 (accessed May 19, 2012).

11 EBR staff writer. 2011. Toshiba, IHI form JV to build steam turbine parts for nuclear plants. Energy Business Review, January 18. http://nuclear.energy-business-review.com/news/toshiba-ihi-form-jv-to-build-steam-turbine-parts-for-nuclear-plants-180111 (accessed May 19, 2012).

12 MAN Turbo. 2008. MAN Turbo to strengthen service presence in the USA. Press release, March 9. http://www.mandieselturbo.com/1013242/Press/Press-Releases/Press-Releases/MAN-Turbo-to-strengthen-service-presence-in-the-USA.html (accessed May 19, 2012).

13 Yahoo! Finance. 2011. Form 10-Q for Capstone Turbine Corp. SEC filings for CPST on February 7. http://biz.yahoo.com/e/110207/cpst10-q.html (accessed May 19, 2012).

14 Hessler, Peter G. 2005. *Power Plant Construction Management: A Survival Guide*. Tulsa, OK: PennWell. P. 121.

15 Maccoby, Michael. 1994. From analyzer to humanizer—raising the level of management thinking. *Research Technology Management*. 37 (5; September/October). Pp. 57–59.

16 *CLP Power Procurement Principles & Practices*. [n.d.] Company brochure.

17 MHI. 2011. MHI Compressor technology licensee in China begins marketing. Press release, February 9. http://www.mhi.co.jp/en/news/story/1102091407.html (accessed May 19, 2012).

18 Doosan Heavy Industries & Construction Co., Ltd. "Doosan Heavy Industries & Construction to Cultivate 200 Suppliers to Global 'Small Giants'" (last modified April 27, 2011).

19 Associated Press. 2009. German conglomerate Siemens buys majority in Yangtze Delta and GIS Steel & Aluminum of China. *Star Tribune*. August 26. http://www.startribune.com/templates/Print_This_Story?sid=55066742 (accessed May 19, 2012).

20 PennEnergy. 2009. India's BHEL signs ten year technology transfer deal with Sheffield Forgemasters. PennEnergy. April 6. http://www.pennenergyequipment.com/article/display.html?id=358365 (accessed May 19, 2012).

21 Dong Energy. 2009. DONG Energy buys A2SEA. Press release, June 25. http://www.dongenergy.com/EN/Investor/releases/Pages/omx%20feed%20list%20details.aspx?omxid=398854 (accessed May 19, 2012).

22 Yamanaka, Megumi. 2009. Toshiba said to buy majority stake in nuclear fuel company. Bloomberg, March 30. http://www.bloomberg.com/apps/news?pid=newsarchive&sid=afRJo8IQYjU4&refer=asia (accessed May 19, 2012).

23 Boston Strategies International analysis, based on a subset of products in electrical distribution and control.

24 General Electric. 2010. GE and Triveni form a joint venture in India to target global power generation market." Press release, April 15. http://www.genewscenter.com/Press-Releases/GE-and-Triveni-form-a-Joint-Venture-in-India-to-Target-Global-Power-Generation-Market-2758.aspx.

25 Green Car Congress. 2011. Mikro Systems and Siemens sign collaborative technology license agreement for gas turbine manufacturing technology; improved turbine efficiency. Green Car Congress, August 19. http://www.greencarcongress.com/2011/08/mikro-20110819.html (accessed May 19, 2012).

26 Zawada, Craig C., Eric V. Roegner, and Michael V. Marn. 2004. *The Price Advantage*. Hoboken, NJ: John Wiley & Sons. P. 258, Appendix 2.

27 Jarboe McFee, Michelle. 2009. Parker Hannifin wins $2.5 billion aerospace contract. Cleveland.com, January 28. http://blog.cleveland.com/business/2009/01/parker_hannifin_wins_25_billio.html (accessed May 19, 2012); Eaton. 2009. Rolls-Royce awards Eaton two applications on Trent XWB aircraft engine. Press release, January 22. http://www.eaton.com/Eaton/

OurCompany/NewsEvents/NewsReleases/CT_202402 (accessed May 15, 2012).

28 Platts Steel Business Briefing. Stalprodukt seals EUR 21m deal with ABB. Lexis Nexis, accessed February 4, 2009. http://www.steelbb.com/us/?PageID=157&article_id=57284.

29 Hessler, 2005.

30 Wood, Allen J., and Bruce F. Wollenberg. 1996. *Power Generation, Operation, and Control.* 2nd ed. New York: John Wiley & Sons.

31 Repsol YPF. [n.d.] ACS 1000 variable speed drive replaces steam turbine in the petrochemical industry. Case study. Application: Blower, 3,000 kW (4,000 hp).

32 Kenyon, Rex. *Process Plant Reliability and Maintenance for Pacesetter Performance.* Tulsa, OK: PennWell.

33 Sen, Dipankar, and Prosenjit Sen. 2005. *RFID For Energy & Utility Industries.* Tulsa, OK: PennWell.

34 Monitran DSP. 2009. Monitran DSP monitors turbine blade conditions. Source. The Engineer, September 7. http://source.theengineer.co.uk/measurement-quality-control-and-test/vision-sound-and-vibration-testing/vibration-sensors/monitran-dsp-monitors-turbine-blade-conditions/352130. article (accessed May 19, 2012).

35 Chin, Will. 2010. GE expands European presence with equity share of Artesis Teknoloji Sistemleri. ARC Advisory Group, September 30. http://www.arcweb.com/asset-lifecycle-management/2010-09-30/ge-expands-european-presence-with-equity-share-of-artesis-teknoloji-sistemleri--1.aspx (accessed May 19, 2012).

36 ABB Group. 2010. ABB factory is voted best in Europe. Press release, September 27. http://www.abb.com/cawp/seitp202/00cb18b719e98a1cc12577ab003ae7f2.aspx (accessed May 19, 2012).

37 Weir Group, The. 2011. "Rapid response programme produces results. *Weir Group Bulletin.* March 2011. http://content.yudu.com/A1rcwe/wb-March-2011/resources/20.htm (accessed May 19, 2012).

CONCLUSION

Successful supply chain management in the oil and gas and power industries relies on four imperatives.

First, the supply chain strategy should be consistent with the economic leverage points of the business. These industries focus heavily on capex and asset reliability, especially power generators and upstream oil and gas companies, because the elements of supply chain management that have the greatest impact relate to capital investment project management, including supply relationships through the acquisition, installation, and commissioning phases. In power utilities, because of the focus on capex, compared to operating costs, adequate attention should be paid to network design and optimization, as well as capital project management. Midstream oil and gas requires analysis of many possible supply chain configurations, and each may involve technical engineering knowledge of shipping, fluid dynamics, or both. Refining activities require a strong management of capital project exposure as experienced through supply chain partnerships. In addition to these areas of particular cost leverage, however, all segments of oil, gas, and power generation offer at least some potential for supply chain improvement through the application of traditional operating and maintenance cost reduction techniques.

Second, a total value chain vision (managing suppliers and their suppliers' suppliers) is required in order to achieve economic targets as well as to ensure safety and environmental stewardship. Demand planning, multi-echelon network modeling, and unbundling analyses require visibility up and down the supply chain. Moreover, rig and pipeline safety requires coordination with suppliers and customers, or else *fault lines*—that is, small misalignments and omissions—will become exposed and could have serious and even disastrous consequences. Therefore, leaders must set a vision and program of action for the extended, multi-enterprise, supply chain.

Third, adequate tools, data, benchmarks, and robust analysis are needed to support good decisions. Last-generation tools are not sufficient, given the financial and safety exposure that comes with commitments of that magnitude. Supply chain practitioners should be thinking about 10 or more years ahead instead of this year or next, using real-options analysis instead of NPV, multilevel integer programming instead of Excel spreadsheets, and custom data sets in addition to data found free on the Web. Long-range scenarios and simulation models should be regularly reviewed and rerun.

Fourth, executive management should commit adequate resources to supply chain management, in consideration of the economic, environmental, and safety risks over which it has a strong influence. Adequately staffed strategic planning and supply chain management departments, risk measurement, and widespread adoption and application of industry standards will pay off in financial and nonfinancial ways.

Hopefully, this book has shed light on the science and art of supply chain management in oil, gas, and power generation by showing how the bits and slivers of supply chain management that are embedded within the organizations, procedures, and processes of the oil, gas, and power generation industries fit with the existing, wider body of functional knowledge of supply chain management.

This book represents merely an initial look at these subjects. Areas for future research and development might include the following:

- Definition of more interoperable and standardized supply chain management processes. To this end, a continued and intensive global effort is needed to improve communication and increase networking opportunities among operators, suppliers, and standards bodies.

- Decision analysis methods that go beyond cost, including such applications as how to make trade-offs between risk and cost, and how to evaluate bundled product-service solutions.

- More—and more specific—processes and procedures related to assuring operational safety in the supply chain, specifically through best practices, standards, and regulations.

BULLWHIP IN THE OIL AND GAS SUPPLY CHAIN—THE COST OF VOLATILITY

Filling an Important Research Gap

Since Jay Forrester described the bullwhip effect in 1958, a vast body of literature has examined how to avoid bullwhip in an industry-agnostic way. This literature has explored, among other topics, the effect of gaming and adjusting order placement mechanisms. For example, Kimbrough, Wu, and Zhong explored how artificial agents can dampen the bullwhip effect, and Ouyang and colleagues tested four different ordering methods (order-up-to, Kanban, generalized Kanban, and order-based).[1]

Many researchers have explored evidence of bullwhip in consumer products and electronics. A consulting study assessed the potential savings from implementation of efficient consumer response in that industry at $30 billion, or 12.5%–25% of cost.[2] Procter and Gamble discovered wide swings in the production and inventory of diapers despite relatively smooth demand; volume discounts and promotional pricing by Italian pasta maker Barilla induced volatility in a system with inherently low end-user demand; HP measured the bullwhip effect by measuring the standard deviation of orders at the stores to the standard deviation of production at the upstream suppliers; and network systems manufacturer Cisco studied the impact of the right to cancel orders on the magnitude of the bullwhip effect.[3]

Few published works have applied system dynamics to the oil and gas industry, and none have quantified the cost of volatility on the upstream oil and gas supply chain. The existing literature in oil and gas has focused on predicting the price of oil rather than analyzing the effect of volatility on in the cost of the oil and gas supply chain. For example, Mashayekhi built a model of oil price drivers, showing that as demand increases, oil price rises, which causes producers to expand capacity, forcing prices down and depressing demand, in a feedback loop.[4] John Sterman came the closest to elaborating the bullwhip effect in the oil and gas industry (specifically with regard to E&P) when he noted that "oil and gas drilling activity fluctuates about three times as much as production."[5] This study shows evidence of the bullwhip effect in the upstream oil and gas supply chain and demonstrates that over time the cost of gasoline is 10% higher as oil producers, oil refiners, heavy equipment suppliers, and their component suppliers pass on the costs of inventory overages and shortages, poorly timed capacity investments, and inflationary prices.

Evidence of the Bullwhip Effect

The pattern of drilling activity and capital investment, by contrast, is characterized by oscillation and amplitude magnification, and it appears that oil price shocks are the root cause. From 1949 until 1973, the average annual price of oil fluctuated within a 7% band, but from 1981 through 2008, the variation leapt to almost 10 times that amount. The 1973 and 1979 oil crises and the sharp escalation and crash of oil prices between 1998 and 2009 introduced a new and seemingly systemic unpredictability to oil prices.

Since the onset of the recent period of oil price volatility that began in 1998, oil drilling investment and activity has tracked the price of oil and has in some cases exaggerated the pattern set by the oil price.[6] The price of oil rose and fell by 52% from its peak of $97/bbl in 1980 to a low of $17/bbl in 1998 and rose again to reach a new high of $97/bbl in 2008.[7] As the price of oil rose and fell between 1998 and 2008, capital expenditure for major oil companies rose and fell by 25%–63%,[8] and the total U.S. rig count (oil and gas) rose and fell by 36%.

Regardless of the reason behind the initial shocks—some think the pattern is cyclical, and there is evidence that it could be chaotic[9]—the variation from a steady-state historical demand clearly induced oscillating and increasing reverberations in production, capacity, and inventory from 1995 to 2009 in markets for oil and gas field machinery and equipment (including turbines and turbine generator sets, motors, electrical equipment, and iron castings). Bullwhip effect is evident in six oil and gas supply markets between 1995 and 2009, as indicated by the measurements in table A–1.

Table A–1. Evidence of bullwhip effect in the oil and gas equipment industry, 1995–2009

Type of Equipment and Metric[a]	Variance	Amplification Ratio[b]	Amplification Difference[c]
Oil & gas machinery and equipment:		5.763	0.0094
Demand	0.002		
Production	0.0114		
Turbines & turbine generator sets:		3.2063	0.0127
Demand	0.0057		
Production	0.0184		
Motors and generators:		2.2201	0.0093
Demand	0.0076		
Production	0.017		
Electrical equipment:		2.275	0.0073
Demand	0.0057		
Production	0.0131		
Casings for machinery:		4.6989	0.0212
Demand	0.0057		
Production	0.027		
Steel fabrications:		3.5495	0.0093
Demand	0.0036		
Production	0.0129		

Note: Bullwhip effect exists when the amplification ratio is greater than one and the amplification difference is positive. [a] Demand is measured by orders received and production (or supplier's production capacity) by units manufactured. [b] Amplification ratio = variance [production] ÷ variance [demand]. [c] Amplification difference = variance [production] – variance [demand].

The swings in capital investment by oil companies caused even bigger swings in the equipment supply chain—in turn causing oscillation in production, inventory, and backlog. While production of turbines and engines declined by 7% between 1998 and 2008, inventories rose by 24%. In an analogous period when new orders spiked three times in 12 years, the backlog of turbine generators tripled and then plummeted to nearly zero twice.[10]

This bullwhip effect causes four types of economic inefficiency at oil companies and their heavy equipment suppliers:

- Oil companies pay higher prices that are set when markets are overheated and are never rolled back when recession hits.

- Equipment manufacturers hold excess inventory during the boom and take a long time to draw it down when the recession hits.

- Equipment manufacturers make excessive capacity investments near the peak and then suffer a low or negative return on investment on it.

- Component and parts suppliers lose orders that they are not able to fulfill at the peak because of inadequate capacity and long lead times caused by large backlogs.

The Cost to E&P Companies, Refiners, OEMs, and Component Suppliers

Boston Strategies International constructed a system dynamics simulation of a four-tier supply chain to test for the presence of—and to quantify the impact of—bullwhip in the upstream oil and gas supply chain. The four tiers were oil producers, oil refiners, heavy equipment suppliers, and their component suppliers (fig. A–1).[11]

The model was used to run two scenarios: a *flat-oil-price scenario* and a *volatile-oil-price scenario*. In the flat-oil-price scenario, an initial shock was simulated, and the aftereffects on the supply chain were traced. In this case, the initial shock was an increase in the price of oil from $30/bbl to $60/bbl. The price of oil rose to a peak of $90/bbl, dropped to $30/bbl, then rose back to $60/bbl to complete a sine wave, with a cycle of 20 years (the whole simulation lasted 43 years). In the volatile-oil-price scenario, after the initial shock, oil price fluctuates in a sine wave with the same overall amplitude as under the smooth-price-cycle scenario but with random oscillation (fig. A–2).

In the simulation, over an extended time frame, the initial increase in demand for oil elevated the production levels of refined oil and drilled crude oil, translating into increased demand for oil field equipment such as oil and gas compressors and turbines. Excess production was found to be higher at the refiner than at the driller, higher at the OEM than at the refiner, and higher at the component supplier than at the OEM during most periods.

Refiners and upstream oil producers pay higher prices that are set when markets are overheated and that are not de-escalated when recession hits. In the simulation, these price hikes added 5% per year to the cost of the equipment, materials, and services that producers buy, after adjusting for inflation caused by metals prices and pure commodity inputs. Moreover, equipment and service prices kept rising even as the price of oil fell, as equipment orders dropped, as capacity utilization dropped by 15%, and as lead times declined to manufacturing throughput time. Capacity adjusted, with a lag, as orders and production fluctuated. The annual cost for refiners was found to be the highest under rough-price scenario during years 8–17.

In the simulation, equipment OEMs incurred high costs in years 11–22 as orders grew owing to rising oil prices, causing equipment OEMs to make excess capacity investments and pay high prices for components as those costs became inflated as well. In fact, prices of turbine hot sections doubled over a 22-year period in the simulation. The capacity additions weighed heavily on the OEMs' finances as orders and backlog declined and bottomed out; ultimately, the OEMs carried that excess capacity for four years too long, although to a lesser degree each year. The OEMs also held excess inventory, which added 8% to the cost of the equipment—similar to the way in which OEMs doubled their inventory between 2004 and 2008, which then became redundant when orders dropped off—and took 12 months to draw down when the recession hit.

Turbine Manufacturing (OEMs)
Total cost for turbine manufacturers

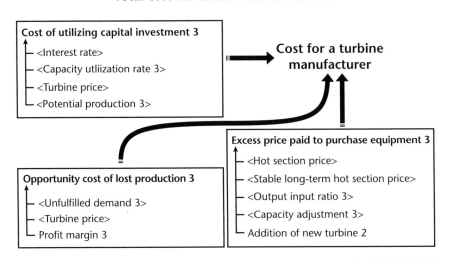

Fig. A–1. Supply chain simulation architecture

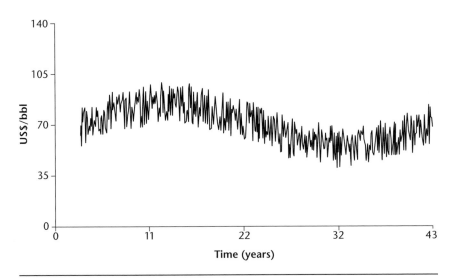

Fig. A–2. Oil price in the volatile-oil-price scenario

Component suppliers lost orders on the upswing and held excess inventory on the downswing in the simulation. Component suppliers were the last ones to see backlogs decline owing to their upstream role in the supply chain, and the approximate halving of their backlog amounted to a depletion of inventory. So, for most of the time during years 11–22, they were depleting inventory. This inventory-carrying cost was their prime supply chain cost. Component and parts suppliers also lost orders that they were not able to fulfill at the peak owing to inadequate capacity and long lead times caused by their large backlogs.

Average annual supply chain costs over a 43-year period in a flat-oil-price scenario totaled $8.3 billion, while in the volatile-oil-price scenario they were $10.3 billion (fig. A–3). The difference, $2 billion, spread across 85 billion b/d, equaled roughly 6.4 cents/bbl. Considering that turbines and compressors represent only 5.8% of oil companies' total external expenditure on equipment, materials, and services, the impact extrapolated to all equipment and services in the oil and gas supply chain was $1.09 (i.e., $0.064/0.058). This is approximately 10% of the weighted average cost of producing a barrel of oil in 2008.[12]

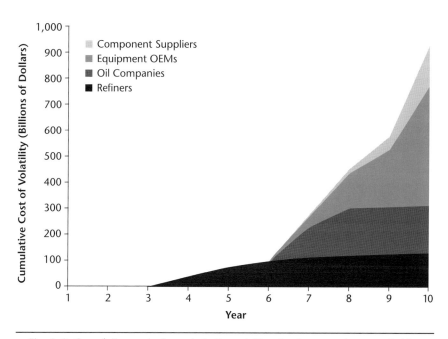

Fig. A–3. Cumulative cost of supply in the volatile-oil-price scenario, years 1–10

Suggestions for Further Research: How to Mitigate the Costs of Bullwhip

A stable environment would help to establish steadier prices and operating profits across the upstream oil and gas supply chain. Stable oil prices and input costs would minimize layoffs during downturns and rehiring during upturns, thereby reducing long-term operating costs. It would also encourage more stable research and development investments, which would result in higher exploration, refining, and distribution productivity because of more rapid and consistent advances in oil and gas equipment technology. This would translate to a higher return on assets.

Therefore, further research on stabilizing strategies would be helpful in determining the course(s) of action that would most effectively mitigate the cost of bullwhip. The research should evaluate the effectiveness of strategies such as (1) long-term supply agreements that are long enough to bridge demand-price-capacity cycles, which can last 20 years or more; (2) sharing of production plans between E&P companies, refiners, OEMs, and component manufacturers; (3) sharing of capital investment; and (4) sharing of supply risk through price indexing and the use of options and futures contracts.

Notes

1 Kimbrough, Steven O., D. J. Wu, and Fang Zhong. 2002. Computers play the beer game: can artificial agents manage supply chains? *Decision Support Systems*. 33. Pp. 1–29. Ouyang, Yanfeng, Alejandro Lago, and Carlos F. Daganzo. 2006. Taming the bullwhip effect: From traffic to supply chains. In *The Bullwhip Effect in Supply Chains*. Ed. Octavio A. Carranzo Torres, and Felipe A. Villegas Morán. New York: Palgrave Macmillan.

2 Poirier, Charles C., and Stephen E. Reiter. 1996. *Supply Chain Optimization: Building the Strongest Total Business Network*. San Francisco: Berrett-Koehler.

3 Lee, H. L., and S. Whang. 2006. The bullwhip effect: A review of field studies. In *The Bullwhip Effect in Supply Chains*. Ed. Octavio A. Carranzo Torres and Felipe A. Villegas Moran. New York: Palgrave Macmillan. Lee, H. L., V. Padmanabhan, and S. Whang. The bullwhip effect: Reflections. In *The Bullwhip Effect in Supply Chains*. Ed. Octavio A. Carranzo Torres and Felipe A. Villegas Morán. New York: Palgrave Macmillan.

4 Mashayekhi, Ali. 2001. Dynamics of oil price in the world market. Paper presented to the 19th International Conference of the System Dynamics Society, July 23–27, Atlanta, GA. http://www.systemdynamics.org/conferences/2001/papers/ (accessed September 10, 2010).

5 Sterman, John. 2006. Operational and behavioral causes of supply chain instability. The In The Bullwhip Effect in Supply Chains. Ed. Octavio A. Carranzo Torres and Felipe A. Villegas Morán. New York: Palgrave Macmillan.

6 Analysis of data from Baker Hughes and EIA.

7 Based on annual data; using monthly data from 1998–2009, the swing was 66%.

8 Chevron: 63%; Royal Dutch Shell: 41%; Exxon Mobil: 26%; BP: 25%.

9 Many factors have contributed to the recent volatility, including political crises, financial speculation, and a sharp increase in demand from developing economies. Two-thirds of respondents to a 2009 Boston Strategies International survey felt that oil prices are caused by speculation by commodity traders and distortions in financial markets.

10 Based on sales of the three largest turbine generator manufacturers between 1948 and 1962. Exhibit material, *Ohio Valley Electric vs. General Electric*, Civil Action 62 Civ. 695, Second U.S. District Court of New York, 1965.

11 Jacoby, David. 2010. The oil price bullwhip: problem, cost, response. *Oil & Gas Journal*. March 22: 20–25.

12 Based on Boston Strategies International 2010 calculations of the all-in cost of purchased materials and services.

COMMON CATEGORIES OF EXTERNALLY PURCHASED EQUIPMENT AND SERVICES FOR THE OIL, GAS, AND POWER INDUSTRIES

1	Exploration
1.1	Seismology
1.2	Reservoir modeling
2	Construction
2.1	Engineering, procurement, and construction services
2.2	Offshore construction and installation
2.3	Onshore construction and installation
2.4	Civil construction
2.4.1	Iron and steel
2.4.2	Nonferrous metals
2.4.3	Rebar
2.4.4	Pipefitting
2.4.5	Steel erection
2.4.6	Steel fabrication
3	Drilling
3.1	Tubulars excluding line and pipe
3.2	Fluids
3.3	Downhole tools
3.4	Bits
3.5	Surface equipment
3.6	Blowout preventers

4.9.2	Casing
4.10	Plug and abandon, decommission
4.11	Contract compression services
4.11.1	Centrifugal compressors
4.11.2	Air compressors
4.11.3	Reciprocating compressors
4.12	Laboratory services
4.13	Production chemical services
4.13.1	Corrosion
4.13.2	Simulation
4.13.3	Sweetening
4.13.4	Hydrate inhibition
4.13.5	Water treatment
4.13.5.1	Field separation systems
4.13.5.2	Macrofiltration equipment
4.13.5.3	Media
4.13.5.4	Membranes and membrane systems
4.13.5.5	Flowmeters
5	Marine services (to support offshore rigs or platforms), excluding very large crude carriers and ultralarge crude carriers
5.1	Mooring
5.2	Tug
5.3	Anchoring
5.4	Crew
6	Offshore
6.1	Foundations
6.1.1	Jacket
6.1.2	Tripod
6.1.3	Tripile
6.1.4	Gravity base foundation

12.3.3   Air-cooled heat exchangers

12.4   Fired heaters

12.5   Piping system

12.5.1   Pipes and valves

12.5.2   Seals and gaskets

12.6   Pumps, compressors, and turbines

12.6.1   Centrifugal pumps for petroleum and gas industries

12.6.2   Positive displacement pumps—reciprocating/controlled volume/rotary

12.6.3   Axial, centrifugal compressors, and expander-compressors

12.6.4   Reciprocating compressors

12.6.5   Rotary-type positive-displacement compressors

12.6.6   Packaged, integrally geared centrifugal air compressors

12.6.7   Steam turbines

12.6.8   Gas turbines

12.7   Steam system

12.8   Cooling-water system

12.9   Other equipment

12.9.1   Vibration, axial-position, and bearing-temperature monitoring systems

12.9.2   Form-wound squirrel-cage induction motors

12.10   Consumables

12.10.1   Chemicals and catalysts

12.10.2   Fuel/electricity

12.10.3   Claus sulfur

12.11   Royalties

12.12   Desalter

12.13   Atmosphere crude still

12.14   Vacuum pipe still

12.15   Coker

GLOSSARY OF TERMS, ACRONYMS, AND ABBREVIATIONS

3PL: third-party logistics service provider. Service provider that manages transport carriers on behalf of a shipper.

4PL: fourth-party logistics service provider. Service provider that manages transport carriers and all other aspects of the logistics function—such as demand and shipment planning, information systems, and inventory management—on behalf of a shipper.

AFE: authority for expenditure.

ANP: National Agency of Petroleum, Natural Gas, and Biofuels (Brazil).

ANSI: American National Standards International. International body that prescribes product and process standards to ensure design quality and consistency and increase interoperability. ANSI is also the parent organization of the electronic interchange that serves as the clearinghouse for U.S. electronic data interchange standards.

API: American Petroleum Institute

ATO: assemble to order. Subset of made to order (see *MTO*) in which standard components are pulled from stock to produce the product once an order is received, and customer or order-specific elements are manufactured and assembled at the last possible point in time (postponement helps to reduce inventory and increase flexibility in production if implemented correctly).

BATNA: best alternative to a negotiated agreement. Course of action that one party will follow if it fails to achieve its desired result in a negotiation. This term was coined by Roger Fisher and William Ury in *Getting to Yes: Negotiating without Giving In.*

bbl: barrel.

Bid Slate Development: The process of developing a short list of qualified suppliers who will be invited to bid on a project.

BOM: bill of materials.

BOO: build, own, operate.

BOOT: build, own, operate, transfer.

bpd: barrels per day.

BSI: Boston Strategies International.

BSI: British Standards Institute.

BTO: build to order. See *MTO*.

Btu: British thermal unit. Unit of energy, approximately equal to 1,055 joules.

bullwhip. Effect of the amplification of changes in inventory or cost upstream in a supply chain, generated by an initial disturbance (often a small change in demand) downstream in the supply chain. Bullwhip is caused by delays in the transmission of information throughout an extended supply chain and is exacerbated by overcorrection, promotions, batching, and tweaking of demand forecasts. Theoretically, it can be eliminated by synchronizing the supply chain, but the degree to which it can be dampened is limited by the length of the supply chain (physical and over time).

capacity reservation. Proactive approach whereby the scheduling system allocates a certain amount of the resources' capacity to attend to specific service types.

capacity utilization. Output as a percentage of possible production at current investment levels.

capex: capital expenditure.

category management. Process that defines the desired end state of procurement activities for each family of externally purchased goods or services.

CCGT: combined-cycle gas turbine.

CCS: carbon capture and sequestration.

CDP: Carbon Disclosure Project. Not-for-profit organization that enlists companies and municipalities to share their carbon footprint data, in an effort to reduce greenhouse gas emissions and encourage sustainable water use.

cfm: cubic feet per minute.

CHP: combined heat and power.

CNG: compressed natural gas.

COGS: cost of goods sold.

concession agreements. Right granted by a government or parastatal organization to a private operator, by which the operator commits to output and reliability levels and agrees to adhere to certain rules.

constraints management. Process of successively identifying the binding constraint, eliminating that constraint, aligning other processes to the new throughput levels, and pursuing the next constraint. (Also called *debottlenecking*.)

critical path. Longest combination of consecutive activities in a project plan. The project duration is constrained by the critical path.

DBO: design, build, operate.

DBOM: design, build, operate, maintain.

DBOO : design, build, own, operate.

debottlenecking. See *constraints management*.

design simplification. Reduction in the complexity that reduces cost, increases performance reliability, or both.

DOE: U.S. Department of Energy.

downstream. Refining (production of gasoline, kerosene, diesel, and other distillate fuel oils, residual fuel oils, and lubricants) and marketing (operation of a retail distribution chain including points of sale).

EMEA: Europe, Middle East, and Africa.

E&P: exploration and production.

e-procurement. Catalogue-based electronic purchasing that may include one or more of the following processes: requisitioning, authorizing, ordering, receiving, and payment.

EOQ: economic order quantity. Algorithm that determines the number of units of an item to be replenished at one time based on minimizing the total costs of acquiring and carrying inventory, which are defined by the unit cost, usage, ordering costs, and inventory carrying cost.

EOR: enhanced oil recovery. Any process for increasing the rate of recovery of hydrocarbon resources from a reservoir after the primary production phase.

EPA: U.S. Environmental Protection Agency. Government agency that establishes regulations pertaining to natural resources (air, water, etc.).

EPC: engineering, procurement, and construction. Contracting arrangement that involves engineering, procurement, and construction, also used to classify a type of company that routinely executes contracts in this mode. EPC firms provide the engineering design, purchase all the equipment and materials needed for the project, and build or oversee the construction.

EPCI: engineering, procurement, construction, and installation. EPC contracting arrangement that also includes installation, which is particularly important in offshore construction, where installation of rigs, platforms, and foundations can be a highly technical and expensive project involving specialized vessels and equipment.

EPFIC: engineering, procurement, fabrication, installation, and commissioning.

ERM: enterprise risk management. Process of identifying, assessing, and managing risks across all the processes and organizational units, to increase sustainability and long-term financial health.

ERP: enterprise resources planning. Companywide information system that centralizes data for increased visibility, accuracy, elimination of redundancy, and rapid access to information that supports decision-making at all levels.

ETO: engineer to order.

EVA: economic value added. A measurement of financial strength that accounts for both cash flows (which are income statement factors) and efficiency of asset utilization (which is balance sheet factors). EVA is defined as net profit after tax less the cost of capital employed. The calculation is as follows: EVA = NOPAT − (capital × cost of capital), where NOPAT is the net operating profit after taxes.

FCC: fluid catalytic cracking.

FEED: front-end engineering and design.

FGSO: floating gas storage and off-loading unit.

FID: final investment decision. Decision on the part of an organization to definitively move forward with a capital project, assuming all the legal and financial risks attached thereto, including commitments to suppliers and supply chain partners. Organizations may budget for a project internally before they reach an FID.

FPSO: floating production, storage, and off-loading vessel.

framework agreement. Stipulation of the terms and conditions that will govern a relationship which both parties expect will consist of multiple contract awards over an extended period of time. While non-Europeans may use the term generically to refer to agreements that cover some but not necessarily all of the terms of subsequent contracts, the European Commission's Directive 2004/18/EC and Directive 2004/17/CE for utilities lay out specific legal requirements for framework agreements as they should be applied in public procurement.

GCC: Gulf Cooperation Council. Group founded in 1981, comprising Saudi Arabia, Kuwait, the United Arab Emirates, Qatar, Bahrain, and Oman. (Also called *Cooperation Council for the Arab States of the Gulf* [CCASG].)

global sourcing. Procurement process involving identifying, pre-screening, and possibly qualifying potential sources in foreign countries, either to lower the cost of externally purchased materials and services by switching suppliers or applying competitive pressure on existing suppliers or to increase quality or technological advantages by introducing new capabilities.

GOSP: gas/oil separation plant; *or*, **gas/oil separation process.**

GTC: gas to commodity.

GTL: gas to liquids.

GTP: gas to power.

GTS: gas to solids.

GW: gigawatt.

HHI: Herfindahl-Hirschman index. Measure of market concentration that is calculated by adding the squares of the market shares of each firm competing in the market. If the HHI is between 1,000 and 1,800, the market is moderately concentrated; if it exceeds 1,800, the market is concentrated.

hp: horsepower.

HPHT: high-pressure/high-temperature.

HSE: health, safety, and environment. Standards, regulations, policies, and organizations intended to protect workers and the environment.

IEA: International Energy Agency.

IEC: International Electrotechnical Commission

IMO: International Maritime Organization.

integrated supply. When a vendor provides resources at the customer site to handle materials and inventory management services for its own and possibly other vendors' products, often in conjunction with a vendor-managed inventory program.

inventory turns. Measure of the velocity of inventory, calculated as the cost of goods sold (COGS) divided by the average inventory level.

IOC: international oil company.

IPP: independent power project.

ISM: Institute for Supply Management.

ISO: International Standards Organization. Group of not-for-profit institutions from 155 countries that develops and publishes standards with the goals of defining international best practices and transcending differences between national standards. Higher-level international standards facilitate and accelerate implementation of best practices worldwide and often eliminate redundancies and wasted resources.

JIT: just-in-time. A manufacturing philosophy that aims to reduce inventory to a one-piece flow involving no buffer stock. JIT is closely related to concepts that enable its execution such as total quality, total productive maintenance, root-cause problem-solving, single-minute exchange of die (fast changeovers), level loading, Kanban, and Kaizen.

joint process improvement. Formal and concerted effort to achieve business benefits by working collaboratively with supply chain partners. Typical benefits are lower overhead or direct costs, shorter cycle times, higher quality levels, and increased reliability and uptime.

Kanban. Tool used in just-in-time (see *JIT*) production whereby cards are attached to replenishment bins or containers to signal the consumption of one unit, thereby triggering a replacement unit. Kanban is often deployed within a batch production environment to trigger production of replenishment parts or components.

kg: kilogram.

KPI: key performance indicator.

KSF: key success factor.

ktpa: thousand tons per annum.

kW: kilowatt.

LCCS: low-cost country sourcing. Procurement strategy to lower costs by switching from high-cost suppliers in mature economies to lower-cost suppliers in emerging economies.

lean management. Set of principles based on delivering what the customer wants without extra steps or waste of any kind. Lean programs such as JIT, Kaizen, and quality management share the same goal, but are distinguished by their focus on values and attitudes, including worker involvement and empowerment, that are needed to sustain continuous improvement in the long run.

lean Six Sigma. Deployment of lean methods, to eliminate waste and increase process efficiency, followed with Six Sigma methods, to reduce process variation.

LEC: levelized cost of electricity. Constant level of revenue necessary to recover all the expenses over the life of a power plant.

life-cycle cost. Total cost of a product, service, or solution from design through disposal, most frequently applied to equipment, in which initial purchase cost needs to be balanced against operating and maintenance costs over the duration of the useful life of the unit(s). Life-cycle cost includes indirect costs, such as the cost of downtime resulting from equipment unreliability, which may result from inferior design or insufficient or improper maintenance. (Also called *total cost of ownership [TCO]*.)

LNG: liquefied natural gas.

LPG: liquefied petroleum gas. (Also called *liquefied propane gas*.)

LSTK: lump sum/turnkey.

LTA: long-term agreement. Contract with a term that is longer than most of the agreements that an organization has historically committed to with suppliers of similar products or services. Many buyers enter into LTAs to achieve lower cost, to secure guarantees, and to obtain better service from their suppliers. Many suppliers enter into LTAs to reduce their long-term cost of sales, to lock in stable revenue streams, and to solidify relationships with strategic customers. The length of contract itself varies widely among organizations since each has a different baseline or historical norm.

LTSA: long-term service agreement. Agreement between a customer and a manufacturer under which the original equipment manufacturer (see *OEM*) is responsible for maintenance, repair, and/or operation of the equipment for a fixed fee. (Also called *contractual services agreement*.)

M&A: mergers and acquisitions.

Macondo. The Macondo crisis refers to the largest marine oil spill in the history of the petroleum industry, which took place April 20–July 15, 2010. It involved an explosion and subsequent fire on the *Deepwater Horizon*, a semisubmersible mobile offshore drilling unit, which was owned and operated by Transocean and was drilling for BP in the Macondo Prospect oil field in the Gulf of Mexico. The explosion resulted in the death of 11 operators and subsequent spillage of 4.9 million barrels of crude oil.

MENA: Middle East and North Africa.

midstream. Transportation and storage of hydrocarbons via pipelines, oil tankers, LNG tankers, and tank farms, including related handling activities.

MMBtu: million Btus.

Monte Carlo simulation. Computer-based repetition of theoretical events, using specific probability distributions, that is designed to provide decision-makers with data on likely outcomes, based on a set of possible choices and stochastic or random influencing variables.

MRO: maintenance, repair, and operating supplies. Parts or components used to support operations and maintenance activities in a manufacturing organization, including consumables and spare parts.

MRP: materials requirements planning.

MSDS: Material Safety Data Sheet. Document that must by law accompany products that contain certain hazardous materials. The MSDS provides information on how to respond if the product spills, if there is a fire, or if people are exposed to it unintentionally. (Also called *PSDS* [*Product Safety Data Sheet*].)

MT: metric ton. (Also called *tonnes*.)

MTBR: mean time between repairs. Interval between planned or unplanned repairs to a piece of equipment. MTBR is thus a measure of operating reliability.

MTO: made to order. Production paradigm whereby production of goods or service does not occur until the organization receives an order from a customer. (Also called *BTO* [*build to order*].)

MTPA: metric tons per annum; *or*, **million tons per annum.**

MW: megawatts.

NACE International. The organization formerly known as National Association of Corrosion Engineers.

network optimization. Process of modifying the configuration of a set of nodes (e.g., plants or warehouses) and flows between the nodes (e.g., inbound or outbound shipments) to minimize the total cost of operating the system (see *life-cycle cost*).

Nm: newton-meter.

NOC: national oil company.

NPV: net present value.

OCTG: oil country tubular goods.

OECD: Organisation for Economic Co-operation and Development.

OEM: original equipment manufacturer. Maker of equipment that sells directly to end customers or through a distributor or a value-added reseller. Distributor and value-added resellers are not OEMs.

OM: operations management. Process of adjusting demand and capacity at work centers throughout a constrained system, to generate output that satisfies customer goals such as cost, quality, and speed. OM is broader than supply chain management (see *SCM*); it applies to both product and service flows, whereas the latter applies primarily to physical flows.

OPEC: Organization of Petroleum Exporting Companies.

operator. Producer of energy.

opex: operating expenditure.

opportunity cost. Cost of an alternative that must be forgone in order to pursue a certain action.

OSHA: U.S. Occupational Safety and Health Administration.

owner. Equity shareholder of an energy producer (in the context of this book).

PAO: polyalphaolefin.

parcel tanker. Tanker ship with multiple holds that are designed to carry many bulk substances, such as chemicals or different grades of petroleum, at one time.

PBL: performance-based logistics. Arrangement under which a logistics provider commits to an agreed level of process performance for a fixed price, possibly including incentives and penalties, rather than conventional billing modes such as time and materials or hourly rates.

power. Any of the following industries: coal-fired thermal power plants (simple cycle: boiler + turbine + generator design); natural gas–fired thermal power plants; cogeneration plants (producing steam for an industrial process, e.g., refining); combined heat and power (CHP); combined cycle (gas turbine generates steam, which is recovered and fed to a steam generator); integrated power and desalination plants; nuclear; hydroelectric; wind; solar; biomass-fueled power plants; municipal solid waste; methane; and electrical distribution & control equipment (including monitoring devices, sensing switches, controls, connectors, motor controls, relays, timers, and temperature controls, programmable logic controllers, and circuit and load protection devices, especially transformers and switchgear).

ppm: parts per million.

PPP: public-private partnership. Agreement between a government entity and one or more private companies to collaboratively build, develop, and/or operate a large-scale facility or network, usually related to power, water, or transportation infrastructure.

predictive maintenance. Type of preventive maintenance that uses statistical analysis and sometimes sensors to predict when required maintenance will need to occur and ensure that maintenance occurs before equipment failure is likely.

price indexation. Practice of linking price changes for a product or service to either raw materials costs or an agreed basket of prices for a given period (typically a month, quarter, or year).

psi: pounds per square inch.

RCRA: Resource Conservation and Recovery Act of 1976.

REACH: Registration, Evaluation and Authorisation of Chemical Substances. European Commission regulation on safe use of chemicals, creating a central database run by the European Chemicals Agency in Helsinki.

real options. Acquisition in the current or a future time frame of tangible assets (e.g., land, property, or equipment) and/or value-added physical transformations (e.g., construction) that offer or increase the flexibility to expand, contract, or change the nature of an investment commitment at a future date. Real options can increase the value of an investment by reducing downside risk. (See also *real-options analysis*.)

real-options analysis. Process of evaluating the benefits and costs of acquiring certain real options.

reliability-centered maintenance. Body of knowledge used to shape the development and optimization of repair and maintenance programs, which involves total productive maintenance, as well as analytical tools common to lean, in order to reduce mean time between repairs (see *MTBR*) and increase uptime.

reverse auction. Electronic auction in which suppliers bid in descending prices and the lowest price wins.

RFID: radiofrequency identification.

RFx: request for information, proposal, or quotation (RFI, RFP, or RFQ, respectively).

ROA: return on assets.

ROHS: Restriction of Hazardous Substances. European Union Directive, implemented in 2006, that prevents companies from marketing new electrical and electronic equipment containing lead, mercury, cadmium, hexavalent chromium, polybrominated biphenyls (PBBs), or polybrominated diphenyl ethers (PBDE) in quantities exceeding specified concentrations.

ROI: return on investment.

RONA: return on net assets.

ROP: rate of penetration.

rpm: revolutions per minute.

SCADA: supervisory control and data acquisition.

SCM: supply chain management. Coordination of the set of activities involved in moving a product (e.g., a machine tool) and its ancillary services (e.g., installation, maintenance, or repair) from the ultimate supplier to the ultimate customer so as to maximize economic value added (see *EVA*). SCM includes manufacturing value added as it accrues along the chain but excludes the initial manufacturing or conversion activity; thus, an initial basic activity such as extraction or farming would be considered a manufacturing activity (i.e., a node), not an SCM activity.

SCOR: supply chain operations reference model.

SEAFOM: Subsea Fiber Optic Monitoring Group.

SECA: Sulphur Emission Control Areas.

should cost. Calculated figure that reflects the price that a supplier could offer if it took full advantage of opportunities for standardization and product simplification, economies of scale from volume purchasing, economies of scope from purchasing multiple types of products or services from the same supplier, and potential savings from better coordination of production and inventory planning with its customer(s).

single sourcing. Intentionally purchasing from only one supplier in order to benefit from close relationships that yield lower cost, higher quality and reliability, and/or shorter lead times through joint problem-solving, than could be obtained with more suppliers.

SKU: stock-keeping unit. Inventoried item with a specific part number.

smart grid. Electricity network that gathers information on usage and production from the behavior of users connected to it and uses that information to better balance supply and demand of electricity to lower cost across the whole network by distributing production to the sources and times when it is most economical and incentivizing consumption from sources and times when it is less costly to produce. Smart grid consists of measurement (meters), networking hardware (transmitters, buses, modems, etc.), data management (SCADA system), and distribution management (relays, controls, demand response software, etc.).

spare parts. Components or subassemblies kept in reserve for use during the repair of equipment. Spare parts must support an irregular demand based on unplanned repairs, as opposed to service parts, which are used to support planned maintenance and thus have a more consistent replenishment requirement.

STOIIP: stock tank oil initially in place.

STP: submerged-turret production.

strategic sourcing. Process for improving the effectiveness of the procurement function. Involves improving spend visibility, understanding supplier capabilities beyond the typical transactional product or service purchase, and developing long-term relationships with strategic suppliers that can help the buyer meet profitability and customer satisfaction goals.

supplier. Contracted provider of services or products.

supply market concentration index. Sum of the squares of the market shares of the suppliers in each category, based on the *Herfindahl-Hirschman index* (see *HHI*) methodology.

supply risk mitigation. Set of coordinated actions that act upon factors that might create disruption or unavailability of materials or services to reduce their likelihood or severity.

system dynamics. Branch of decision science that relies on modeling of feedback loops and delays in complex systems in order to identify and reduce the incidence of uneconomical resource allocation and to improve system performance as measured by target output levels such as cost, flexibility, and service level.

TCO: total cost of ownership. See *life-cycle cost.*

T&D: transmission and distribution.

third-party maintenance. Maintenance provided by a vendor other than the original equipment manufacturer (see *OEM*) or the user. See *3PL.*

T&M: time and materials.

TMS: transportation management system. Computerized system to manage fleet activities, especially by assigning loads to the lowest-cost qualified carrier. TMS also may include functionality to decide on modes of transport, manage export-import paperwork, and optimize load planning.

tons: short tons.

tonnes: metric tons.

tpa: tons per annum. (Also called *tons per year* [*tpy*].)

TPM: total productive maintenance.

tpy: tons per year. (Also called *tons per annum* [*tpa*].)

TQM: total quality management. Engagement of the workforce in ensuring processes that continuously eradicate quality problems through the use of process analysis tools.

TSCA: Toxic Substances Control Act of 1976.

UAV: unmanned aerial vehicle.

ULCC: ultralarge crude carrier.

upstream. Exploration and production of crude oil and natural gas; drilling, completing, and equipping wells; and processing (e.g., separation, emulsion breaking, desilting, and field petroleum gathering, including shale plays and oil sands).

vertical integration. Acquiring a supplier in order to assure supply or enter into a new or related line of business.

VLCC: very large crude carrier.

VMI: vendor-managed inventory. Mode of inventory management in which a supplier monitors and regulates the amount of inventory at a customer's location to keep supply and demand in balance.

WMS: warehouse management system.

WTI: West Texas Intermediate crude oil.

WTO: World Trade Organization.

Appendix D

ADDITIONAL RESOURCES

Ackermann, Thomas (ed.). 2005. *Wind Power in Power Systems*. 2nd ed. Hoboken, NJ: John Wiley & Sons.

Agência Nacional do Petróleo. 2011. Local content framework, the Brazilian experience. Rio de Janeiro, February 6. Powerpoint document by Marcelo Mafra Borges de Macedo. http://www. google.com/url?sa=t&rct=j&q=&esrc=s&frm=1&source=w eb&cd=1&ved=0CEsQFjAA&url=http%3A%2F%2Fwww. britcham.com.br%2Fdownload%2F020611rj_marcelo_mafra. pdf&ei=2JnLT_vVEui60QGCnZmrAQ&usg=AFQjCNGpLY eGHK7-D9krOqmtaPMayVjEvA.

Alvarado, Vladimir, and Eduardo Mannrique. 2010. *Enhanced Oil Recovery: Field Planning and Development Strategies*, Burlington, MA: Gulf Professional.

Anderson, Roger N., Albert Boulanger, John A. Johnson, and Arthur Kressner. 2008. *Computer-Aided Lean Management for the Energy Industry*. Tulsa, OK: PennWell.

Association of Energy Engineers. 2002. *Integrated Solutions for Energy and Facility Management*. Liburn, GA: Fairmont Press.

August, Jim. 2000. *Applied Reliability-Centered Maintenance*. Tulsa, OK: PennWell.

Barnett, Dave, and Kirk Bjornsgaard. 2000. *Electric Power Generation: A Nontechnical Guide*. Tulsa, OK: PennWell.

Bartnik, Ryszard, and Zbigniew Buryn. 2011. *Conversion of Coal-Fired Power Plants to Cogeneration and Combined-Cycle: Thermal and Economic Effectiveness*. New York: Springer.

Baxter, Richard. 2005. *Energy Storage: A Nontechnical Guide*. Tulsa, OK: PennWell.

Bloch, Heinz P., and John J. Hoefner. 1996. *Reciprocating Compressors: Operation and Maintenance*. Boca Raton, FL: Gulf Professional.

Burdick, Donald L. 2001. *Petrochemicals in Nontechnical Language*, Tulsa, OK: PennWell.

Bureau of Economic Analysis and the National Association of Manufacturers. 2005. *The Facts about Modern Manufacturing*, 7th ed. Washington DC.

CAPS Research, and Institute for Supply Management. 2011. *Petroleum Industry Supply Management Performance Benchmarking Report*. Tempe, AZ. P. 8.

Carson, Richard T., Michael B. Conaway, W. Michael Hanemann, Jon A. Krosnick, Robert C. Mitchell, and Stanley Presser. 2004. *Valuing Oil Spill Prevention: A Case Study of California's Central Coast*. Norwell, MA: Kluwer Academic.

Centre for Economics and Management (IFP-School). 2004. *Oil and Gas Exploration and Production: Reserves, Costs, Contracts*. Paris: Editions Technip.

Çerçi, Yunus 2010. *The Thermodynamic and Economic Efficiencies of Desalination Plants: Performance Evaluation of Desalination Plants*. Saabrucken, Germany: Lambert Academic Publishing (Lap Lambert).

Chambers, Ann. 1999. *Natural Gas & Electric Power in Nontechnical Language*. Tulsa, OK: PennWell.

———. 1999. *Power Primer: A Nontechnical Guide from Generation to End-Use*. Tulsa, OK: PennWell.

Chambers, Ann, and Barry Schnoor. 2011. *Distributed Generation: A Nontechnical Guide*. Tulsa, OK: PennWell.

Christopher, Martin. 2005. *Logistics and Supply Chain Management: Creating Value-Added Networks*. 3rd ed. Harlow, UK: Pearson Education.

Cipollina, Andrea, and Giorgio Micale. 2009. *Seawater Desalination: Conventional and Renewable Energy Processes*. New York: Springer.

Committee of Sponsoring Organizations of the Treadway Commission. Enterprise Risk Management—Integrated Framework Executive Summary. 2004. http://www.coso.org/guidance.htm.

Commonwealth of Australia. 2011. "National Alliance Contracting Guidelines." Department of Infrastructure and Transport. http://www.infrastructure.gov.au/infrastructure/nacg/index. aspx.

Cory, Jordan A. 2010. *Combined Heat and Power: Analysis of Various Markets*. New York: Nova Science.

Cox, Andrew and Michael Townsend. 1998. *Strategic Procurement in Construction*, London: Thomas Telford.

Degeare, Joe P., and David Haughton. 2003. *Gulf Drilling Guides: Oilwell Fishing Operations: Tools, Techniques, and Rules of Thumb*. Boca Raton, FL: Gulf Professional.

Delea, Frank, and Jack Casazza. 2010. *Understanding Electric Power Systems: An Overview of the Technology, the Marketplace, and Government Regulation*. 2nd ed. Hoboken, NJ: John Wiley & Sons.

De Renzo, D. J. (ed.). 1983. *Cogeneration Technology and Economics for the Process Industries*. Park Ridge, NJ: Noyes Data.

Devereux, Steve. 1999. *Drilling Technology in Nontechnical Language*. Tulsa, OK: PennWell.

Devine, Michael D. 1987. *Cogeneration and Decentralized Electricity Production: Technology, Economics, and Policy*. Boulder, CO: Westview Press.

Dickson, Mary H., and Mario Fanelli (eds.). 2003. *Geothermal Energy: Utilization and Technology*. Sterling, VA: Earthscan.

Drinan, Joanne E. 2001. *Water and Wastewater Treatment: A Guide for the Nonengineering Professional*. Boca Raton, FL: CRC Press.

Duval, Marc. 2010. Offshore activities and supply chain markets: FPSO market. Paper presented at the 5th Annual Global Procurement and Supply Chain Management for the Oil and Gas Industry, Barcelona, September 26–28, 2011.

Ebenhack, Ben. 1995. *A Nontechnical Guide to Energy Resources: Availability, Use, and Impact*. Tulsa, OK: PennWell.

Emadi, Ali, Abdolhosein Nasiri, and Stoyan B. Bekiarov. 2005. *Uninterruptible Power Supplies and Active Filters*. Boca Raton, FL: CRC Press.

European Wind Energy Association. 2009. *Wind Energy—the Facts: A Guide to the Technology, Economics and Future of Wind Power.* Strerling, VA: Earthscan.

Fingas, Mervin (ed.). 2011. *Oil Spill Science and Technology: Prevention, Response, and Cleanup.* Burlington, MA: Elsevier.

Flynn, Damian (ed.). 2003. *Thermal Power Plant Simulation and Control.* Power and Energy Series, vol. 43. London: Institution of Engineering and Technology.

Forsthoffer, William E. 2005. *Forsthoffer's Rotating Equipment Handbooks: Fundamentals of Rotating Equipment.* New York: Elsevier.

Forsund, Finn R. 2010. *Hydropower Economics.* New York: Springer.

Gary, James H., and Glenn E. Handweck 2001. *Petroleum Refining: Technology and Economics.* 4th ed. New York: Marcel Dekker.

Gerwick, Ben C., Jr. 2007. *Construction of Marine and Offshore Structures.* 3rd ed. Boca Raton, FL: CRC Press

Grace, Robert D. 2003. *Blowout and Well Control Handbook.* Burlington, MA: Elsevier.

Horlock, J. H. 1996. *Cogeneration—Combined Heat and Power: Thermodynamics and Economics.* Malabar, FL: Krieger.

Hughes, Joanne B. (ed.) 2009. *Manhole Inspection and Rehabilitation.* 2nd ed. Reston, VA: American Society of Civil Engineers.

International Energy Agency. 2007. *Fossil Fuel-Fired Power Generation: Case Studies of Recently Constructed Coal- and Gas-Fired Power Plants.* Cleaner Fossil Fuels. Paris: International Energy Agency.

International Standards Organization. 2008. *Code of Practice for Risk Management.* Draft BS 31100.

Jacoby, David. 2004. On the cutting edge of strategic sourcing: how industry leaders are planning to transform supply relationships over the next four years. Boston Strategies International. http://www.bostonstrategies.com/images/BSI_-_SS1_On_the_Cutting_Edge.pdf.

———. 2005. The new economics of partnering. *Purchasing.* January 13.

———. 2005. "High-Impact Sourcing: Precision-Guided Strategies for Maximum Results." White Paper. http://www.bostonstrategies.com/images/BSI_-_SS2_High-Impact_Sourcing.pdf.

———. 2007. Budgeting for volatility. Presentation to Demand Planning & Forecasting Best Practices Conference, New Orleans. April 27. Institute of Business Forecasting. Page 32.)

———. 2007. "Supply Risk Management." Webcast. November 19. http://www.bostonstrategies.com/images/Supply_Risk_Management_080319_EXCERPT_110613_dj.pdf.

———. 2009. *Guide to Supply Chain Management*. New York: Bloomberg.

———. 2010. The oil price bullwhip: problem, cost, response. *Oil & Gas Journal*. March 22: pp. 20–25.

———. 2010. Balancing economic risks: tips for a well-structured deal. *Middle East Energy*. September. Tulsa: Pennwell. Pp. 5–7.

———. 2011. Global trade restrictions and related compliance issues pertaining to oil and gas production chemicals. Society of Petroleum Engineers technical paper # OTC 22005-PP, May 2011. Prepared for presentation at the Offshore Technology Conference, Houston, May 2–5.

———. 2011. Uncovering economic and supply chain success in the new emerging economies. Paper presented to APICS International Conference in Pittsburgh, October 24.

Jacoby, David, and Figueiredo, Bruna. The Art of High-Cost Country Sourcing. Supply Chain Management Review, June 2008. Pages 36).

Jahn, Frank. 1998. *Hydrocarbon Exploration and Production*. Developments in Petroleum Science Series, vol. 46. Amsterdam: Elsevier Science.

Jeffs, E. 2008. *Generating Power at High Efficiency: Combined Cycle Technology for Sustainable Energy Production*. CRC Press (Taylor & Francis, London) and Woodhead Publishing (Cambridge), UK.

Jorb, W., and B. H. Joergensen. 2003. *Decentralised Power Generation in the Liberalised EU Energy Markets*. New York: Springer.

Kaimeh, Philip. 2003. *Power Generation Handbook: Selection, Applications, Operation, Maintenance*. New York: McGraw-Hill.

Kaltschmitt, Martin, and Wolfgang Streicher. 2007. *Renewable Energy: Technology, Economics and Environment.* New York: Springer.

Kehlhofer, Rolf, and Bert Rukes. 2009. *Combined-Cycle Gas & Steam Turbine Power Plants.* 3rd ed. Tulsa, OK: PennWell.

Lee, H. L ., V. Padmanabhan, and S. Whang. 1997. Information distortion in a supply chain: The bullwhip effect. *Management Science.* 32. April. Pp. 546–558.

Leffler, William L. 2000. *Petroleum Refining in Nontechnical Language,* Tulsa, OK: PennWell.

Leffler, William L., and Gordon Sterling. 2003. *Deepwater Petroleum Exploration & Production: A Nontechnical Guide.* Tulsa, OK: PennWell.

Lorange, Peter. 2005. *Shipping Company Strategies: Global Management under Turbulent Conditions.* Amsterdam: Elsevier.

Lyons, William. 2010. *Working Guide to Drilling Equipment and Operations.* Boca Raton, FL: Gulf Professional.

Macnab, Alistair M. 2010. *The Fundamentals of BreakBulk Shipping: A Primer and Refresher Study for Global Logistics Students and Professional Logisticians, Ashore and Afloat.* Boston: Pearson Learning Solutions.

Mariani, E., and S. S. Murthy. 1997. *Control of Modern Integrated Power Systems.* Advances in Industrial Control. New York: Springer.

Marsh, W. D. 1980. *Economics of Electric Utility Power Generation.* New York: Oxford University Press.

McCrae, Hugh. 2003. *Marine Riser Systems and Subsea Blowout Preventers.* Austin, TX: The University of Texas at Austin.

Miesner, Thomas O., and William L. Leffler. 2006. Oil and Gas *Pipelines in Nontechnical Language.* Tulsa, OK: PennWell.

Mokhatab, Saeid, and William A. Poe, and James G. Speight. 2006. *Handbook of Natural Gas Transmission and Processing.* Burlington, MA: Elsevier.

Myers, Philip. 1997. *Aboveground Storage Tanks.* New York: McGraw-Hill.

Najafi, Mohammad. 2010. *Trenchless Technology Piping: Installation and Inspection.* New York: McGraw-Hill.

Nardone, Paul J. 2009. *Well Testing Project Management: Onshore and Offshore Operations.* Boca Raton, FL: Gulf Professional.

Naujoks, Boris, and Theodor J. Stewart. 2010. *Multiple Criteria Decision Making for Sustainable Energy and Transportation Systems.* New York: Springer.

Northcote-Green, James, and Robert G. Wilson. 2007. *Control and Automation of Electrical Power Distribution Systems.* Boca Raton, FL: CRC Press.

Osphey, Cedrick N. 2009. *Wind Power: Technology, Economics and Policies.* New York: Nova Science.

Page, John S. 1977. *Cost Estimating Manual for Pipelines and Marine Structures.* Houston: Gulf Professional.

Pavithran, K. V. 2004. *Economics of Power Generation, Transmission, and Distribution: A Case Study of Kerala.* New Delhi: Serials Pub.

Pehnt, Martin, and Martin Cames. 2010. *Micro Cogeneration: Towards Decentralized Energy Systems.* New York: Springer.

Petchers, Neil. 2003. *Combined Heating, Cooling & Power Handbook: Technologies & Applications: An Integrated Approach to Energy Resource Optimization.* Liburn, GA: Fairmont Press.

Porter, Michael E. 1980. *General Electric vs. Westinghouse in Large Turbine Generators (A).* Boston: Harvard Business School.

Rajan, G. G. 2006. *Practical Energy Efficiency Optimization.* Tulsa, OK: PennWell.

Read, Colin 2011. *BP and the Macondo Spill: The Complete Story.* New York: Palgrave Macmillan.

Rebennack, Steffen, and Panos M. Pardalos. 2010. *Handbook of Power Systems I.* New York: Springer.

———. 2010. *Handbook of Power Systems II.* New York: Springer.

Rustebakke, Homer M. 1983. *Electric Utility Systems and Practice.* New York: John Wiley & Sons.

Santorella, Gary. 2011. *Lean Culture for the Construction Industry: Building Responsible and Committed Project Teams.* New York: Productivity Press.

Sathyajith, Mathew. 2006. *Wind Energy: Fundamentals, Resource Analysis and Economics*. New York: Springer.

Scheer, Hermann. 2004. *The Solar Economy: Renewable Energy for a Sustainable Global Future*. Sterling, VA: Earthscan.

Sears, S. Keoki, Gleen A. Sears, and Richard H. Clough. 2008. *Construction Project Management: A Practical Guide to Field Construction Management*. 5th ed. Hoboken, NJ: John Wiley & Sons.

Seba, Richard D. 2003. *Economics of Worldwide Petroleum Production*, Tulsa, OK: PennWell.

Seddon, Duncan. 2006. *Gas Usage & Value*. Tulsa, OK: PennWell.

Seger, Karl A. 2003. *Utility Security: The New Paradigm*. Tulsa, OK: PennWell.

Shively, Bob, and John Ferrare. 2011. *Understanding Today's Natural Gas Business*. Laporte, CO: Enerdynamics.

———. 2010. *Understanding Today's Electricity Business*. Laporte, CO: Enerdynamics.

Shively, Bob, John Ferrare, and Belinda Petty. 2010. *Understanding Today's Global LNG Business*. Laporte, CO: Enerdynamics.

Sirchis, J. (ed.). 1990. *Combined Production of Heat and Power*. London: Elsevier Applied Science.

Slater, Philip. 2010. *Smart Inventory Solutions: Improving the Management of Engineering Materials and Spare Parts*. New York: Industrial Press.

Society of International Gas Tanker and Terminal Operators. 2008. *LNG Shipping Suggested Competency Standards: Guidance and Suggested Best Practice for the LNG Industry in the 21st Century*. 2nd ed. London: Witherby Seamanship International.

Spellman, Frank R. 2009. *Handbook of Water and Wastewater Treatment Plant Operations*. Boca Raton, FL: CRC Press.

Spitzer, David W., and Walt Boyes. 2003. *The Consumer Guide to Magnetic Flowmeters*. Chestnut Ridge, NY: Copperhill and Pointer.

Spliethoff, Hartmut. 2010. *Power Generation from Solid Fuels*. Power Systems. New York: Springer.

Tessmer, Raymond G., and John R. Boyle. 1995. *Cogeneration and Wheeling of Electric Power: Opportunities in a Changing Market.* Tulsa, OK: PennWell. 1995

Tiratsoo, John. 1999. *Pipeline Pigging and Inspection Technology.* Houston: Gulf Professional.

Today's trends: Offshore rig construction costs. 2010. *Rigzone.* February 15. http://www.rigzone.com/news/article. asp?a_id=87487.

Tusiani, Michael D., and Gordon Shearer. 2007. *LNG: A Nontechnical Guide.* Tulsa, OK: PennWell.

Underhill, Tim. 1996. *Strategic Alliances: Managing the Supply Chain.* Tulsa, OK: PennWell.

Urboniene, Irena A. 2010. *Desalination: Methods, Costs and Technology.* New York: Nova Science.

U.S. Department of Labor Bureau of Labor Statistics. 2012. *PPI Detailed Report.* April. P. 11.

Venem, Jan Erik. 2007. *Offshore Risk Assessment: Principles, Modelling and Applications of QRA Studies.* 2nd ed. Springer Series in Reliability Engineering). London: Springer.

Warkentin-Glenn, Denise. 2006. *Electric Power Industry in Nontechnical Language.* Tulsa, OK: PennWell. 2006

Whittick, J, and D. Edmonds. 1993. *Petroleum and Marine Technology Information Guide: A Bibliographic Sourcebook and Directory of Services.* Edited by A. Myers, J. London: E & FN Spon.

Wilson, Jeff. 2004. *The Model Railroader's Guide to Industries Along the Tracks.* Waukesha, WI: Kalmbach.

Wright, Charlotte J., and Rebecca A. Gallun. 2008. *Fundamentals of Oil & Gas Accounting.* 5th ed. Tulsa, OK: PennWell.

INDEX

logistics cost reduction in,
107–108
management of oil, 182–
183
of natural gas, 171–173
of oversized project cargo,
162–163
Traverse Drilling, 164
Tungsten, xviii
turbine manufacturing, 232
Turnbull Report, 119
TWP Projects, 116
Tyco, 154

U

unionization, 109
Unipetrol, 195
United States
drilling and production in,
xv
manufacturing prices in, 43
pipeline safety bill in, 174
refineries in, 189
upfront costs, 11
upstream activities, 4
capital procurement in, 132
cost containment in, 7
oil and gas in, 133, 146
usage-based maintenance, 91

V

Valero Energy, 122
value chain, 4, 5, 225
value-added services, 48, 110
value-creation triangle, 144–
145
vendor-managed inventory
(VMI), 102–104, 220
vertical integration, 41, 212
very-long-term contracts, 136–
137

Vickrey auctions, 84
volatile-oil-price scenario,
230–233
volume purchasing, 154

W

warehousing
information technology
managing, 181–182
logistics cost reduction in,
107–108
system for, 103
waste elimination programs,
106
waste water, 126
water resource depletion, xviii
waterfall chart, 155
Weatherford, 60, 164–165
WellDynamics, 72
well-logging company, 60
West Africa, 81
Westinghouse Electric, 213, 214
wind power industry, 37
Wood Group, 164
work structure risks, 26
work-in-process (WIP), 104
Workplace Safety Rule, 159
Wu, D. J., 227

Z

zero defects, 94